ONE MAN'S ISLAND

KEITH BROCKIE

ONE MAN'S ISLAND

A NATURALIST'S YEAR

1817

HARPER & ROW, PUBLISHERS, New York

Cambridge, Philadelphia, San Francisco,
London, Mexico City, São Paulo, Sydney

Printed in Italy by
Arnoldo Mondadori Editore, Milan, for
Harper & Row, Publishers
10 East 53rd Street
New York, N.Y. 10022

Library of Congress Cataloging in Publication Data

Brockie, Keith
  One man's island.

      1. Natural history——Scotland——May, Isle of.
2. Seasons——Scotland——May, Isle of.   3. May, Isle of
(Scotland)   I. Title.
QH141.B76   1984   508.412'9   84–47708
  ISBN 0–06–015360–1

CONTENTS

1 Breeding Birds

2 Other Wildlife

3 Migration

4 Grey Seals

Legend

P	Pencil
PC	Pencil Crayon
WC	Water Colour (vis White Body Colour)
CP	Conte Pencil
C	Conte
B	Biro
O	Oil

Inset map (upper right):

10 miles · 10 Kms

FIFE

Crail · Fife Ness · Anstruther · Caiplie · Methil · Elie · Isle of May

Kirkcaldy · Firth of Forth · Fidra · Lamb · Craigleith · Bass Rock

Inchcolm · Inchkeith · North Berwick

Inchmickery

Forth Bridges

Edinburgh · LOTHIAN · Dunbar

N
W — E
S

Main map labels:

Norman Rock
Shag Rock
Mars Rocks
Silver Sands
North Ness
Whaup Rest
North Horn
Rona
West Tarbet
Standing Head
Nybo (Iron) Bridge
East Tarbet
Burnett's Leap
Altarstanes (West Landing)
Tarbet Hole
Horse Hole
St Andrew's Well
Clett
West Head
North Plateau
High Road
Low Light
The Bield
The Middens
Loose Tooth
McLeod's Path
Burrian
Colm's Hole
Three-tarn Nick
Holyman's Road
Bishop Cove (and Stack)
Peregrine's Nest
Beacon
'Island' Wreck
Island Rocks
Top Trap
Tower
H
Arnott Trap
Foreigner's Point
The Spout
Greengates
Mill Door
Craigdhu
Palpitation Brae
Byres
Kettle
Dam
Loch
Fluke St
Ringing Hut
Kettle Ness
Black Haugh
Low Trap
Mouse House
Greer Face
High Tarn
Bain (S.O.C.) Trap
South Plateau
Jetty
Kirkhaven (East Landing)
Thistle Field
The Pillow
Cornerstone
Chapel
The Buss
Dense
Tennis Court
Low Ledge
Lady's Cave
Well
Cross Park
Gully Trap
Ardcarron
The Angel
South Ness
The Pilgrim
South Horn
Beacon
Pilgrims' Haven
Maiden's Bed
Swirrey
Maiden Hair
Cleaver

Legend:

⬤ Vegetation
⬤ Bare rock above high water mark
⬤ Rock exposed at low tide
🦅 Heligoland Traps

The Isle of May

0 — 1 Kilometre
½ mile

Low Light   Oct 83

INTRODUCTION

The rising sun silhouettes the puffins standing on top of the rocks, while hundreds more whirr overhead on their tiny wings. Lesser black-backed gulls wheel in alarm above my head, annoyed at this bleary-eyed intruder emerging into the dawn from the building known as the Low Light; the gulls' nearest nest is not thirty yards from the front door. At the top of the tower I can see a rock pipit, its beak full of flies – food for its newly fledged brood voraciously calling from nearby locations. All around me, echoing across the sea, are the soft cooing-calls of the eider drakes displaying to the ducks. Fulmars soar effortlessly over the busy kittiwake colony whose nests are situated on precarious cliff ledges. The dew glistens on the flowering clumps of pink thrift that blanket the surrounding area along with the white sea campion. The sea crashes continually on the rocks below, and a gentle breeze wafts the pungent aroma of the seabird cliffs up into the atmosphere. It is a feast of sight, sound and smell – this early summer day on the Isle of May, my home and workplace for most of 1983.

The Isle of May lies at the mouth of the Firth of Forth, Scotland, just over five miles south-east of the villages of Crail and Anstruther in Fife and ten miles north-east of North Berwick in the Lothian region. It is just over a mile long on an axis running north-west to south-east and half a mile at its widest point. Its area is approximately 140 acres and, with a tidal fall of about fifteen feet, a considerable extra area of rock is exposed at low tide.

The high western cliffs rise to 150 feet and a tilted plateau slopes to a low rocky shore on the east coast. All the shoreline is rocky except for a boulder beach at Pilgrims' Haven and two broken beaches of shell sand at Kirkhaven and Silver Sands. Along the West Cliffs the sea has carved a spectacular series of rock stacks, arches and caves. The less scenic east coast is broken by many small inlets and offshore reefs.

Pilgrims' Haven is particularly beautiful, marred only by the flotsam, especially plastic rubbish, which is washed up by the tide. It is dominated by the Angel Stack which rises vertically almost 100 feet from the sea, and is backed by sheer cliffs. The stones in Pilgrims' Haven are rounded by the constant turmoil of the sea. Another magnificent rock stack stands in Bishop Cove. Mill Door is a large natural arch, easily accessible, from which the cliffs of Greengates can be viewed.

A number of faults traversing from east to west divide the island into four parts at high tide – the North Ness, Rona, the May proper and the Maiden Rocks. The top of the island is covered by soil of varying thickness, generally composed of dry crumbly peat or sandy loam, and there are traces here of raised beach material. The climate is equable with less rain than on the Scottish mainland – although the spring of 1983 had far too much of its share of rain and fog!

The island, owned by the Northern Lighthouse Board, was declared a National Nature Reserve in June 1956, under an agreement with the Nature Conservancy Council, in order to safeguard its internationally important colonies of seabirds. The island has one of the highest densities of seabirds in the North Sea. These seabirds are, mainly in order of abundance, puffin, guillemot, kittiwake, herring gull, shag, lesser black-backed gull, razorbill, eider duck and fulmar. The N.C.C. provides a warden to help both with research and with visitors during the summer months, and more recently it has also provided a warden for some of the winter months. The island is famous for its Bird Observatory and Field station – the first to be founded in Scotland for the study of migration – and it was set up here because the prominent position of the May off the coast of Scotland makes it the first landfall for many wayward migrants during the spring and autumn.

The history of the Isle of May has been well documented and dates as far back as the middle of the twelfth century. Much of the following has been extracted from *The Isle of May* by W. J. Eggeling (1960), the most comprehensive account of the island and its wildlife (now unfortunately out of print). In the twelfth century, during the reign of King David I, a monastery was founded on the island dedicated to St Adrian, an Irish missionary martyred here. Adrian and his followers were based in the caves at Caiplie between Cellardyke and Crail on the Fife coast. After fleeing the mainland Adrian was killed on the May (about the year 875) by marauding Danes who were devastating the country from their longboats. In the Middle Ages the island became a place of pilgrimage with visits from royalty. King James IV in particular was a frequent visitor. In June 1508 he arrived on the island with a shooting party and an entry in the Accounts of the Lord High Treasurer of Scotland states that the sum of sixteen pence was paid to 'ane rowbote that hed the king about the Isle of Maii to schut at fowlis with the culveryn'. Presumably King James was shooting at the auks, kittiwake and eider.

During the 1715 Jacobite rising the Earl of Strathmore and some 300 men sought temporary refuge on the May. They kept the pursuing English ships at bay and managed to escape under cover of darkness after a few days. Later the island supported a small smuggling and fishing community. The caves of the May provided some splendid hiding places. One particularly cunning hiding place, shored up by wood and concealed by shingle,

was discovered some years ago in Kirkhaven. In 1814 ten acres of the island were enclosed for the cultivation of barley, potatoes, turnips, carrots and other vegetables. Livestock, including sheep, cattle and goats, was kept on the island at various times. During the World Wars this century the island was garrisoned with the Royal Observer Corps, although it was never actually attacked and their lookout has now been demolished. The most striking monument that still stands on the island is thus the medieval Chapel, its walls now colonised by thrift and other plants and used as a nest site by rock pipits and pied wagtails.

By the main lighthouse stands a white-washed 'keep-like' building. This is the lower half of the beacon – the first lighthouse ever to be built in Scotland. Originally a two-storey building some 40 feet high, erected in 1636, it was in use for 150 years. For fuel the beacon used coal, which was hauled up the tower by a rope and pulley windlass and burnt in a large raised grate surrounded by a parapet. A ton of coal was consumed nightly, three tons on a very windy night. To pay for this, a levy was placed on every ship using the mouth of the Forth, double the amount for foreign (including English) ships. In 1810 representations were made to get a more dependable light after two frigates mistook a limekiln on the Lothian coast for the beacon and were wrecked on the shore. The new light, the Tower, came into use in February 1816 when the beacon was reduced to a single storey.

The Low Light, in turn, came into use in the autumn of 1844. It was a stationary light so that ships leaving the Tay estuary could navigate past the Carr Rocks off Fifeness by sighting the position of its beam in relation to the revolving light of the Tower. The Low Light was made redundant at the end of the century when the North Carr Lightship was anchored off the Carr Rocks. Today four keepers man the Tower light, with its diesel generators and fog horns. The keepers are relieved at monthly intervals. Up until 1972 their families also stayed with them but now that the May is classified as a rock station the keepers' families are not allowed to remain on the island.

In the past 100 years some thirty-nine ships have come to grief on the island and its reefs. One of these, the *Island*, a Danish vessel of 1774 tons, still forms a landmark on the rocks near Colm's Hole where, in dense fog, it was wrecked on 13 April 1937. Ironically, this was the skipper's last voyage before he retired after 260 runs plying between Iceland, Copenhagen and Leith. Happily all 66 people on board were rescued by the Anstruther lifeboat. Most of the cargo was taken off in the next few days, and on 18 April 1937 the inhabitants of the Observatory boarded the ship at low tide and procured food, crockery and bedding which would otherwise have gone to waste. Overcoming their lack of knowledge of Danish, palatable bottles and cans of food were liberated and consumed with relish during the following weeks. The daily log entries from that year describe these events vividly, and crockery with the Danish Seaways imprint is still in use in the Low Light! Today the ship's tangled, rusty remains lie well above the high-tide mark, testimony to the awesome power of the sea.

The Bird Observatory was founded in September 1934 by members of the Midlothian Ornithological Club. It was the second British bird observatory to be established (the first was established on the island of Skokholm off the coast of Pembrokeshire, Wales, a year earlier, in 1933, by R. M. Lockley). Inspired by the work done on migration on the island of May by two ladies, Misses Baxter and Rintoul, between 1907 and 1933, plans were made to establish a trapping station. The Observatory was housed in the old coastguard house and the first trap was erected with the help of W. B. Alexander and R. M. Lockley. Apart from a gap during the war years the Observatory has continued to operate ever since. The Low Light has been its headquarters since 1946, courtesy of the Northern Lighthouse Board. The observers resident in the Low Light maintain a migration log, a daily census and special note books for descriptions of any rare birds seen. Qualified ringers are expected to work the Heligoland traps and use the rings provided in the ringing hut (see the Migration section). The Low Light houses the 'Chatty Logs', a series of fourteen thick volumes to date, chronicling the daily happenings in an often witty manner. This makes fascinating reading during the evening in front of the driftwood fire.

The Low Light basically consists of a living room with two beds in an alcove, a bedroom with four beds, a kitchen, a washroom and an outside chemical toilet. Water comes from a spring, the St Andrew's Well, on the hill above the Observatory. The living room is heated by a wood-burning stove (the shoreline provides ample driftwood), there is a calor gas cooker, and light comes from gas and paraffin lamps. All in all it is a spartan but comfortable 'bothy' well in accord with the island.

I first visited the island in September 1973, with members of the Tay Ringing Group, essentially to catch migrants. During that stay I did some sketches of the wildlife, including a late shag's nest with chicks, which were included in the Observatory log. I wince every time I look back at them but it's nice to go over the logs from year to year to see my artistic progression. Since that first visit I have been to the May many times, including three separate Christmas weeks, and the island was home to me for nearly a year while I collated sketches for this book.

In order to paint with conviction a wildlife artist can only really get to know so much in his own lifetime. Working for so long on the island – not just through 1983, but over the last decade – I have begun to feel part of it, absorbing its atmosphere and feeling at one with the environment and wildlife. I had always wanted to spend a long, concentrated period on the island and the year that was necessary to produce the sketches for this book proved the ideal way of satisfying that ambition. Moreover, from my own point of view, living in Perthshire as I do, the Isle of May has many advantages, such as its relative ease of access in comparison to many other islands, the approachable nature and variety of its wildlife in all seasons and a comfortable base in the Low Light for working during inclement weather.

I hope this book will show the reader a small part of the tremendous variety of wildlife which the island, despite its small size, has to offer. Every season brings its surprises. Originally I was going to divide the book into four chapters depicting life during the seasons but this proved to be impossible because of the overlap between them. For instance, breeding starts in late March when the first shag eggs are laid and some birds still have chicks in the nest well into October. The feral pigeons have an even more extended breeding season. Migration occurs, to some extent, in virtually every month of the year. Consequently I decided to change the projected structure of the book and so the first section now deals with the most important aspect of the island – its breeding birds. I have tried to show features of their display and breeding biology, from eggs to chicks, and also, in the painting of the guillemot ledge (31–2), to give an impression of a large colony. The second section shows some of the other forms of wildlife on May, except for the birds and seals. The choice here was vast, from the seashore life to the plateau vegetation, so I have simply drawn the flora and fauna that I personally find the most interesting. Migration provided the material for the third section. The vast journeys undertaken by some of these small migrant birds, always at the mercy of the weather, are staggering; and the thrill of seeing an unexpected rare bird is part of the island's lure. My fourth and final section is on the grey seals which abound in the sea and which I love drawing; it is fascinating to contrast their lovely sleek lines and form when wet with their awkward 'blancmange-like' shape on land. The sketches also show the development of seal pups from birth to independence.

The weather rules my life on the island, dictating the amount of field-work I can do. I dread the resounding blasts of the fog-horn (South Horn has four short blasts with the North Horn sounding a long blast 67 seconds later) for it means that my activities will swiftly be curtailed. A solid week of fog in the month of May remains vividly etched in my memory. Visibility was down to less than fifty yards with a penetrating dampness that made drawing even more difficult. This happened at the busiest time of my year. Indeed I lost a great many days' work in the spring of 1983 because of the excessive fog and rain. Wind is a constant companion, though it is usually only a problem when it is up to gale force and lashes salt spray over the island. Otherwise I can normally find a sheltered spot with an interesting subject in view. The cold weather in early spring, late autumn and winter invariably restricts the time I can spend in exposed locations, forcing me to draw quickly. In fact, this is good training and, given practice, one can economise and get the maximum of information from the minimum of time. The coloured paper I use, incidentally, prevents any glare from the sun reflecting off the paper. I then annotate the drawings with 'scrawled' notes and colour codes giving desirable information which I work up as soon as possible, back in the Low Light, with the subject still fresh in my mind.

For convenience I carry all my drawing equipment in a multi-pocketed fisherman's vest. There is a large zipped pocket on the back which accomodates my small drawing board (9 × 13 in) plus a folder with assorted coloured paper. I also carry a piece of foam rubber for a cushion – sitting on rock for a while can get painful! This jacket keeps everything handy for a quick departure in case something exciting turns up and I want to rush off to sketch it. I also use a larger hollow drawing board some 20 × 24 × 1 ins in dimension. This is constructed of thin plywood bordered by 2 cm-square wood sides, one of which is hinged for access. This dual-purpose board is very useful. Firstly, I use it for larger drawings and for stretched watercolour paper, the carrying strap doubling for use as a neck strap to support the

board if I am standing sketching. Secondly, the hollow space allows the safe transit of my drawings, which are also sealed in a plastic bag, during the boat journeys between the island and mainland.

The majority of my field sketching is drawn via a telescope mounted on a tripod which leaves my hands free to hold the board and sketch. The telescope I use is an Optolyth 30 × 75, its optical brilliance and large object lens being particularly advantageous. The 30-times magnification is powerful enough for my purposes; any bigger and the image tends to get darker and more distorted by heat haze. The use of a telescope distances me from the subject allowing a more relaxed attitude in the creature than if I was sitting near enough to cause it alarm. Drawing relaxed and sleeping birds also allows me to get good details of bills, wings, proportions etc. These details facilitate the sketching of more lively birds which seldom remain inactive. Measured reference drawings from the freshly dead birds which are occasionally found are of great value. Some sketches take a lot longer than others, especially when I am trying to draw a particular aspect of behaviour. Patience is then called for while I get the details pencilled in bit by bit – provided the behaviour is repeated, such as during the 'cackling' display of the fulmar.

There are a few golden rules to follow for any wildlife artist, primarily an awareness and sense of wonder at the wildlife around one. The learning process is never-ending. Observation and knowledge of the subject will pay dividends time and time again. No wildlife has a definitive form or colour; individuals of a species vary as much as humans do. It is always wise to question your ideals and methods so as to prevent your work 'standing still'. It is also worth trying out new techniques and materials, and gleaning ideas from other artists' work. I get much pleasure browsing through books depicting the work of other wildlife artists such as Bruno Liljefors, Joseph Crawhall, Charles Tunnicliffe, Eric Ennion, John Busby, Gunnar Brusewitz and Lars Jonsson. In July 1982 I visited Läckö Slott in Västergötlands, Sweden, to see a massive exhibition of the work of the old Swedish master Bruno Liljefors (1860–1939). His gargantuan canvases with powerful impressionistic renderings of wildlife subjects were awe-inspiring. Indeed it was seeing those paintings that inspired me to the challenge of painting the large oil of the guillemot ledge.

I feel very privileged to have been able to work full-time for nearly a year on a place I love, doing what I love. But the title of this book, *One Man's Island*, is really a misnomer. The island belongs to the wildlife, not to me, and the National Nature Reserve agreement will safeguard this with as little human intervention as possible. And that's how it should be.

Keith Brockie
Inchture, January 1984

Acknowledgments

In the preparation of this book I have received help from many people. First and foremost I would like to thank Mike Harris and Sarah Wanless for their help during the year. Mike, a scientist from the Institute of Terrestrial Ecology, has been studying the auks on the island since 1972. Sarah was the Nature Conservancy summer warden for 1983. My thanks must also go to the wardens of N.C.C. South-East Region and the lighthouse keepers of the Northern Lighthouse Board, to the Isle of May Bird Observatory committee, particularly Bernard Zonfrillo, to the Sea Mammals Research Unit, David Pullan (N.C.C. winter warden), Alison Beck (for use of the aerial photograph), the many observers staying in the Low Light during the year, and to the boatman Jimmy Smith for his expert seamanship in often difficult conditions. Finally a special thanks to Morag for her encouragement and support.

Access and Accommodation

Access to the Isle of May is available by boat from Crail and Anstruther for day visitors during the summer months – numbers, tide and weather permitting. Boats can be chartered through local fishermen for larger organised parties. For further information on the island contact the Regional Officer, N.C.C., 12 Hope Terrace, Edinburgh EH9 2AS.

Accommodation is limited to parties of six, usually weekly, in the Low Light at a current (1984) nightly charge of £2 per person. This is run by the Isle of May Bird Observatory Committee and preference is given to birdwatchers during the spring and autumn.

Hon. Sec.
Bernard Zonfrillo,
28, Brodie Road,
Glasgow G21 3SB.

Bookings Sec.
Mrs Rosemary Cowper,
9, Oxgangs Road,
Edinburgh EH10 7BG.

1
BREEDING BIRDS

The Isle of May is internationally important for its large colonies of breeding seabirds. It is one of the best documented islands in Britain and much research has been carried out on it, especially in the last decade.

During 1983 the breeding bird population on the island included 1,855 occupied shag nests, 22,600 individual guillemots, 2,300 razorbills, over 11,000 occupied puffin burrows, 106 occupied fulmar sites, 545 eider duck nests (105 ducklings were reared to fledgling stage – a good year!), eight pairs of shellduck (four nest burrows located), 2,578 herring gull nests, 1,385 lesser black-backed gull nests, over 6,000 kittiwake nests and 37 common tern clutches belonging to at least 29 pairs (some re-lays). Thirty pairs of oystercatcher and four pairs of lapwing nested, though the latter were unsuccessful in their attempts to breed because of the predations of gulls and crows. In addition the breeding passerines included two pairs of stock dove nesting in burrows, about 30 pairs of feral pigeon, 10–20 pairs of starling, 25–30 pairs of rock pipit, one pair of meadow pipit, three pairs of pied wagtail and three pairs of swallow with only three broods raised from six clutches. Other species which have bred in recent years include the great black-backed gull, Arctic tern, ringed plover, carrion crow, song thrush, blackbird and linnet.

In recent years most of the seabirds on the May have been increasing at an annual rate of between 5% and 22%. It is thought that this is because of the preponderance of small fish in the surrounding sea due to the over-exploitation of large fish by commercial fisheries. Natural disasters, however, sometimes take their toll of the breeding seabirds. In 1976 the shag population dropped to 365 pairs, the result of a heavy mortality of adults due to a 'red tide' (a paralytic shellfish poisoning caused by algae). Their numbers, however, have now recovered to an all-time high of 1,855 occupied nests. Severe weather is another hazard. At the end of the first week in February 1983 at least 20,000 auks were washed up dead on beaches from the Firth of Forth south to the Thames estuary near London. This was caused by severe north-easterly gales and a shortage of sprats, leading respectively to exhaustion and starvation. Very few auks turned up on the May, fortunately, although birds originating from here were involved.

Some 42 ringed puffins and smaller numbers of ringed guillemot and razor-bill from the May were recovered during this disaster, the worst of its kind ever recorded in Britain. The volume of tanker traffic to the oil terminals further up the Forth means that oil-pollution is also an ever-present threat. Some birds were oiled in February 1983, but on a small scale. The consequences of a large oil spill in the Forth do not bear thinking about.

We still know relatively little about the wintering areas of many of our breeding seabirds, but recoveries of the auks killed in the wreck indicated not only that the majority of birds were from Scottish colonies but that birds from Norway and Iceland were also involved. Without ringing, such data would be impossible to obtain. As well as its scientific value, ringing seabirds is very therapeutic to me after long hours sitting drawing; I need some physical effort, such as abseiling down cliffs to ring kittiwake chicks (I ringed over 800 in 1983), to counteract the mental exertions.

There are so many different kinds of bird in the breeding colonies that I am spoilt for choice when it comes to drawing them. The majority are very tame and approachable making my work so much easier; I don't have to spend hours searching for a suitable subject to sketch. Indeed, sitting on top of a cliff with thousands of birds in view, it is often hard to know where to start!

Shags are great fun to draw, craning their snake-like necks as they watch me, their huge spatulate webbed feet gripping the slippery rocks. From January till April they sport a crest which enhances their comical looks. Most people don't see them in their full splendour as the crest is quickly lost as soon as they start incubating their clutch. Most of the shags are very confiding and one can sit alongside the nest drawing the bird without causing much disturbance. Here the artist can score over the photographer in depicting the lovely green and bronze sheen of the adult shag's plumage. Most photographers 'lose' this colouring because they try to compensate for the glaring white background of the guano. From the defiant display of head shaking and calling one can differentiate between the sexes of the nesting shags. The males emit a guttural croak whilst the females can only manage a plaintive hissing noise. Their nests are a large jumble of seaweed, sticks and any other flotsam which takes their fancy. Amongst other things, I have found plastic dolls, a camera cable shutter release, hacksaw blades and tin cans tucked into their nests. Some shags seem to have a predilection for taking objects with certain colours such as blue and red. Unguarded nests will quickly be ransacked by neighbouring shags. I have even seen some shags rummaging about in the outhouses of the Low Light looking for nesting material. Their young are really grotesque, particularly when small and naked. One male, nicknamed Philandra and recognised by colour rings, is a real character. He has lost countless eggs through fights with other males and conquests of new females. This year he actually managed to raise a youngster to fledgling for the first time, but unfortunately the chick died later. Probably it was born too late in the season to survive.

The puffins are another success story on the island. In the past twenty years the number of breeding puffins has increased markedly from around 500 birds in 1963 to over 11,000 occupied burrows in 1983. This increase took place initially as a result of large immigration from other colonies, notably the Farne Islands, as is indeed proved by ringing recoveries. Puffins are comical-looking birds with their waddling gait and large brightly coloured bill. The bill is well adapted to carrying, crossways, a large number of fish. A lateral hinge allows the mandibles to exert equal pressure from the gape to the bill tip. In addition the roof of the upper mandible is serrated to help grip the slippery fish. Sand eels are their main prey and the incoming puffins laden with fish often have to run the gauntlet of hungry gulls. Some of the gulls are very adept at catching the puffins in flight and robbing them of their meal before they can dive down their burrows. The chicks leave their burrows at roughly six weeks of age to make their way down to the sea. This takes place usually under cover of darkness in order to avoid the depredations of the gulls. Some of the chicks find their way into the lighthouse diesel generator building down in what is called Fluke Street. They probably mistake the throbbing of the generators for the noise of the sea. At times, especially in the evening, great hordes of puffins circle round their colonies like massive swarms of bees.

Watching the guillemot chicks leaving the cliffs at dusk is an extraordinary sight. I sometimes go and sit down by Mill Door, where the imposing cliffs of Greengates are lit by a fiery red and yellow sunset, and where I am surrounded by the constant clamour of the auks and kittiwakes. High above me a guillemot chick is about to launch itself off the cliff. At around 20 days old it is only a third of its adult size, yet it parachutes off the ledge, accompanied by the adult, and falls some 120 feet to the sea. It flutters down on its tiny underdeveloped wings, splashes into the sea and immediately dives and swims expertly, calling out all the while to maintain contact with the parent. The darkness affords some protection from marauding gulls. The chick then swims out to sea with only the male parent in attendance. Time and time again I have watched these plucky youngsters leaping off the cliffs – it is an impressive baptism. From the cliff-tops on a calm day one can watch some of the adults swimming under the water, often playfully torpedoing another floating guillemot upon surfacing. They use their wings for propulsion underwater and their backs take on a shimmering silver colour as a result of air bubbles that are trapped under their feathers.

Whereas the guillemots favour broad, exposed ledges, the razorbills prefer crevices and corners of ledges. Thus their eggs are not so pear-shaped (pyriform) as those of the guillemots and are less susceptible to rolling off. They are smart-looking birds much enhanced by a thin white line between the eye and bill with a further stripe vertically on the bill. When they yawn they reveal a bright yellow interior to their bill and throat which contrasts vividly with their black and white plumage.

In recent years the number of eider ducks on the May has been increasing, and over 500 nests were found in 1983. They usually nest in vegetation or by a rock, but I decided to draw some of the more unusual locations in which they nested, such as the nest under the foghorn pipe and the one I actually found in a gull's nest. Some eider ducks nest in what seem to be very unlikely places: one female nested under a set of wooden steps by the Tower and never flinched at all when people climbed up and down only inches above her head. This trusting nature makes eider ducks ideal models for drawing. They usually lay four to six eggs, although more than one female can lay eggs in the same nest. The nests are insulated with down which the female plucks from her underparts. In 1983 the eiders managed to rear about 150 chicks to the fledgling stage which is very good in comparison to previous years. Many nests are lost to gull predation; this is often the result of visitors who, sometimes unwittingly, disturb the nests and subsequently fail to cover the eiders' eggs with down to conceal them. Gulls also predate the chicks before the female can lead them to the relative safety of the sea. The male eiders are very handsome birds in their black and white plumage with a salmon-pink flush to the breast and green patches on the nape. Particularly in April and May the seashore echoes with their soft cooing as groups of males call and display to the females, throwing back their heads and puffing out their breasts.

The fulmar is a comparative newcomer to the Isle of May, the first breeding being proved in 1930. The island of St Kilda was the only breeding colony in Britain and Ireland up until 1878 but, since then, fulmars have begun to increase in number and they now breed all round the British coastline. There are over a hundred occupied nesting sites on the May, mostly on the West Cliffs, although the fulmars are gradually colonising other areas on the island. They are difficult to draw mainly because of their complicated beak structure and tubular nostrils, typical of the petrel family. Most of the island's adult fulmars are quite shy but by proceeding stealthily one can usually find a few 'tame' individuals to get close to. Taking care not to startle fulmars is essential as the adults and especially the chicks have an unpleasant habit of ejecting fish oil from their stomach at intruders. This defence mechanism is surprisingly accurate at distances up to three feet. Fulmar oil also has a foul smell which is extremely difficult to wash out of clothes completely. Sometimes they spit oil on other seabirds which get too close to their nests. I found a guillemot plastered with fulmar oil washed up in Pilgrims' Haven. I also have the wings of a young male golden eagle found near the Aberdeenshire coast which was so disabled by matted fulmar oil that it could barely fly and soon starved to death.

The gull population used to be much higher than at present with about 15,000 pairs of herring gulls and 2,500 pairs of lesser black-backed gulls. The island was almost becoming a gull-slum, so quickly were the numbers increasing. The denudation of vegetation by the gulls was also causing a great deal of soil-erosion: areas such as the Maidens were once clad in turf instead of the bare rock found there today. So in 1972 a culling programme was started to control the number of breeding gulls. Numbers have now been reduced to around 4,000 pairs, with remaining nests destroyed to prevent recruitment in later years. Thankfully, the vegetation is now recovering. This form of gull-control is now also practised on many islands around Britain.

The lesser black-backed gulls are summer visitors to the May from March to October, as the many ringing recoveries prove. They winter as far south as Portugal and Morocco. Some individuals of this species can be very aggressive in defence of their chicks, giving a hefty clout to the head of the unwary. The greater black-backed gull has also bred in small numbers during the past decade.

The kittiwakes are beautiful, delicate gulls compared to their larger cousins. They derive their name from their cry 'kittiwaak'. Their calls are the most attractive of all the May's seabirds. I love to sit on top of the cliffs on a calm evening listening to their varied calls ranging from a laughing 'ha-ha-ha' to their soft 'cooing' sounds. Their breeding season starts in late February when the first adults return to their nest ledges to hold territory. They nest in spectacular locations on the precipitous cliffs, the nests cemented onto small ledges with their guano. They also use mud collected from pools to compact the nesting material. Huge flocks can descend on the pools taking beak-fulls of mud and grass back to the cliffs. Kittiwakes lay two to three eggs which are then incubated by both parents for around 25 days. After hatching, the chicks take about six weeks to fledge. Upon fledging they are fully independent of the adults. The young are very striking, especially in flight, displaying a black 'M' running from wingtip to wingtip with a black collar and a black tip to the tail. The kittiwakes are increasing in numbers at roughly 5% a year and new colonies are springing up on the slopes around the Loch and on Rona. One enterprising pair nested successfully in 1982/3 on a small cliff by the busy path along Holyman's Road.

Vast numbers of terns used to breed on the Isle of May. In 1946 the tern population was at its height with 5,000–6,000 pairs of common tern, 400–500 pairs of Arctic tern, 1,400–1,500 pairs of Sandwich tern and 15–20 pairs of roseate tern. The demise of the tern colony has been blamed on the gulls taking over their nesting areas but terns are in fact very fickle birds and one cannot predict their numbers from year to year. Indeed the terns are now making a comeback after many years' absence and at least 29 pairs of common tern bred on the North Plateau in 1983. Accurate numbers are difficult to obtain because of the deep vegetation in which they breed; and one does not want to disturb them unnecessarily. The terns on the island will, it is hoped, continue to increase as more colonies desert the mainland in the face of pressure from human expansion. Fife has lost many such colonies, especially on the sandy shores of Tentsmuir.

As a challenge and a contrast to the small sketches and paintings in this book I decided to try and paint a large oil (6 × 4 feet) featuring a densly packed guillemot ledge (31–2). My aim was to give an idea of a bustling sea-bird colony – all but the smell that is! 'Dense', an aptly named study colony on a broad ledge overlooked by a hide, provided an ideal subject. The hide both enabled me to sketch the birds through a telescope without disturbing them and also shaded the paper from strong sunlight. Most illustrations and photographs show guillemots in the idealised, bolt-upright postures of alert birds. I wanted a composition showing birds preening, fighting, sheltering and feeding chicks, and sleeping and soaking up the sun, so strong sunlight was essential to help bring out the form of the birds, for their relatively uncomplicated two-colour plumage lends itself to shadow. This would allow me to indulge in a broader rendering without having to go into much plumage detail. First I drew a grid plan of the painting area, delineating the rock structure, and identified each square by a reference letter and number. Then, through many hours of observation, I built up a series of sketches of groups and individual birds which I marked with the grid code referring to the actual square. Later I drew the birds onto the grid on a smaller scale to build up a 'cartoon' composition of the ledge. This I transferred to the canvas by drawing it up square by square on the larger scale.

In order to capture the same light and shadow conditions sketching could really only be done on sunny days between 11 am and 1 pm. For extra reference I painted detail sketches of guillemot chicks and eggs. I also kept in a freezer any freshly dead adults which I found. These I used later for colour references and for details of feet and wings. For head and bill details I sketched a 'bridled' bird from different angles. The 'bridled' is a variant form of the guillemot with a white eye-ring and stripe. Between 3% and 5% of the May's guillemots are bridled and this proportion increases the further north the colony is found.

People have asked me, 'Why not just use a photograph of the colony, it would be so much easier?' Maybe, but a photograph can only capture an instant whereas a painting is a distillation of many hours of observation. And as a result of embarking on this oil painting I learnt a lot about the behaviour of guillemots, the play of light on their plumage, and so on. I hope that the outcome, even after scaling it down to fit the size of this book, will satisfy and interest the reader as much as it did me.

preening its back feathers.

preening its head and
neck feathers using the
'comb' on one of its claw
nails

immature (1st year) shag, South Ness 26th February 83

some details from a dead bird from East Tarbet.

1

details from a Shag found freshly
dead in Pilgrims Haven, prominent
crest on the crown (on both sexes)
present from December to May

31ˢᵗ January - 2ⁿᵈ February 83.

1/1

1/1

1/1

'comb' on this claw
for preening

large spatulate webbed feet
set well to the rear of the
body for underwater swimming

2

very low tide due to the full
moon, the kelp fronds in the
background are not normally
uncovered.

both watching me

3 adult Shags sketched via a telescope near the Swiney, South Ness, 1st March 83

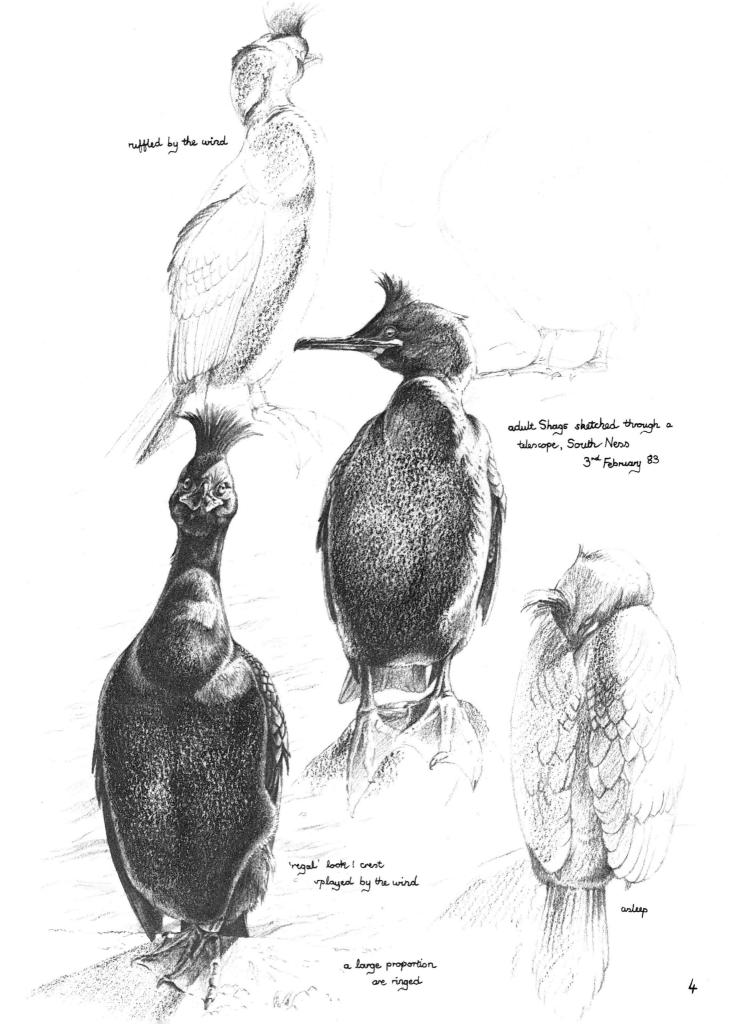

ruffled by the wind

adult Shags sketched through a
telescope, South Ness
3rd February 83

'regal' look! crest
vplayed by the wind

a large proportion
are ringed

asleep

4

Shags incubating their eggs, South Ness
2nd April 83

This year the first eggs were
noted on the 24th March

an albino and a
normal Herring Gull chick

Ardcarran 17th June

the albino survived till the 30th
but was not seen after then

Kittiwake chick, Low Light
14th June.

bald & reptilian looking Shag chick, @ 6 days old
14th June 83

parting in the sun

Shags with large chicks
below the South Horn, 16th Aug 83

a fly which
got in the way!

Shags with their chick, 29-30th August, a late
nest on Rona by the Iron Bridge.
– on the evening of the 4th Sept a westerly gale up to force 9
washed this nest and others off the cliff face.

8

Puffins on top of the air raid
shelter by Holyman's Road,
23rd July 83

9

yawning

sleeping Puffins
'Burian', 11ᵗʰ July 83

no strong shadows

10

Puffins, Bishop's Cove   18ᵗʰ July 83

some have
extensive pale
markings on their
backs

preening

11

2-3 yrs old

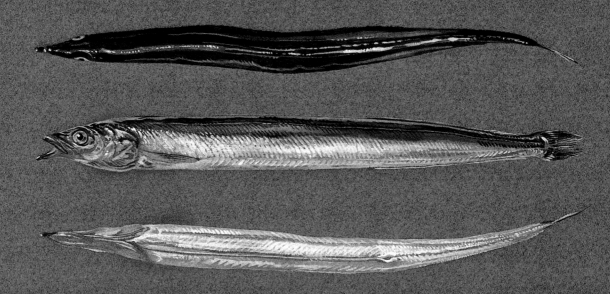

Sand Eels ⅟₁ 20ᵗʰ June 83

dropped by a puffin, these are their main prey item

⅟₁

young Puffin chick which I
took out of a burrow to sketch
for a few minutes.

still with down on it's head & neck

¹⁄₁

Juvenile Puffin @ 7 weeks old

Holyman's Road , 9ᵗʰ July 83

At this age the young Puffins leave their burrows
usually under the cover of darkness to minimise
gull predation. They walk and tumble into the
sea and head far out, plucky little birds!

the inside claws on puffin's feet
are flattened, possibly as an aid
to digging burrows?

14

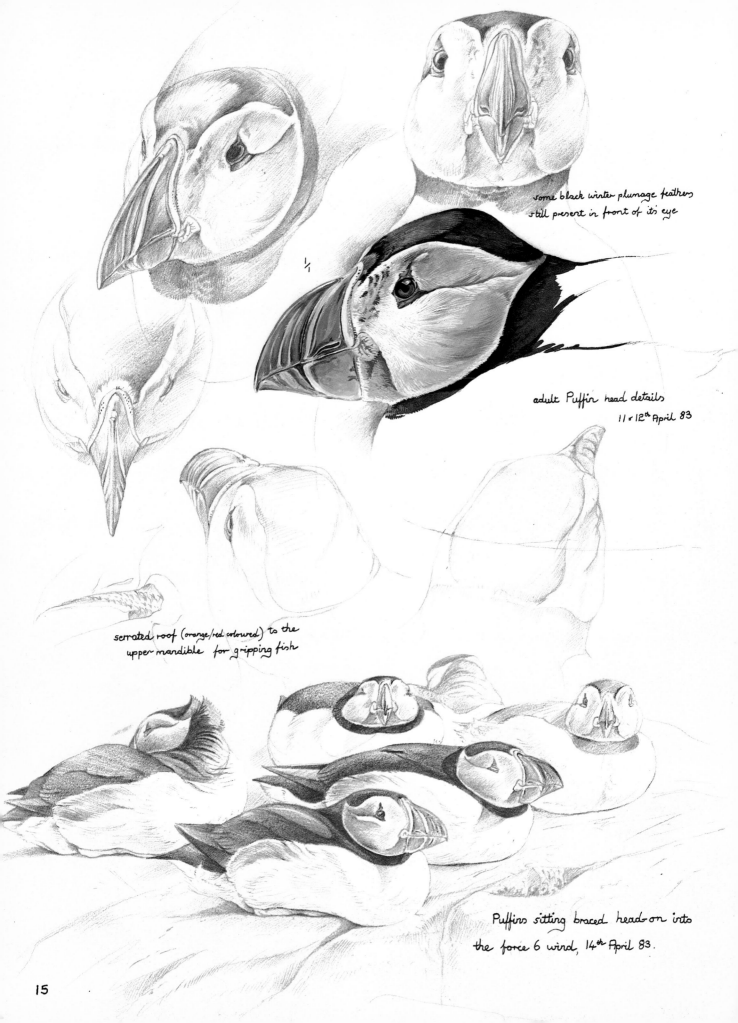

some black winter plumage feathers
still present in front of its eye

$\frac{1}{1}$

adult Puffin head details
11 & 12th April 83

serrated roof (orange/red coloured) to the
upper mandible for gripping fish

Puffins sitting braced head-on into
the force 6 wind, 14th April 83.

sleeping & soaking up the sun

ruffled incubating bird
in partial shadow

Razorbills
Mill Door 16th July 83

16

under the parents' wing!

Razorbills with chicks
Bishop's Cove 5ᵗʰ July 83

17

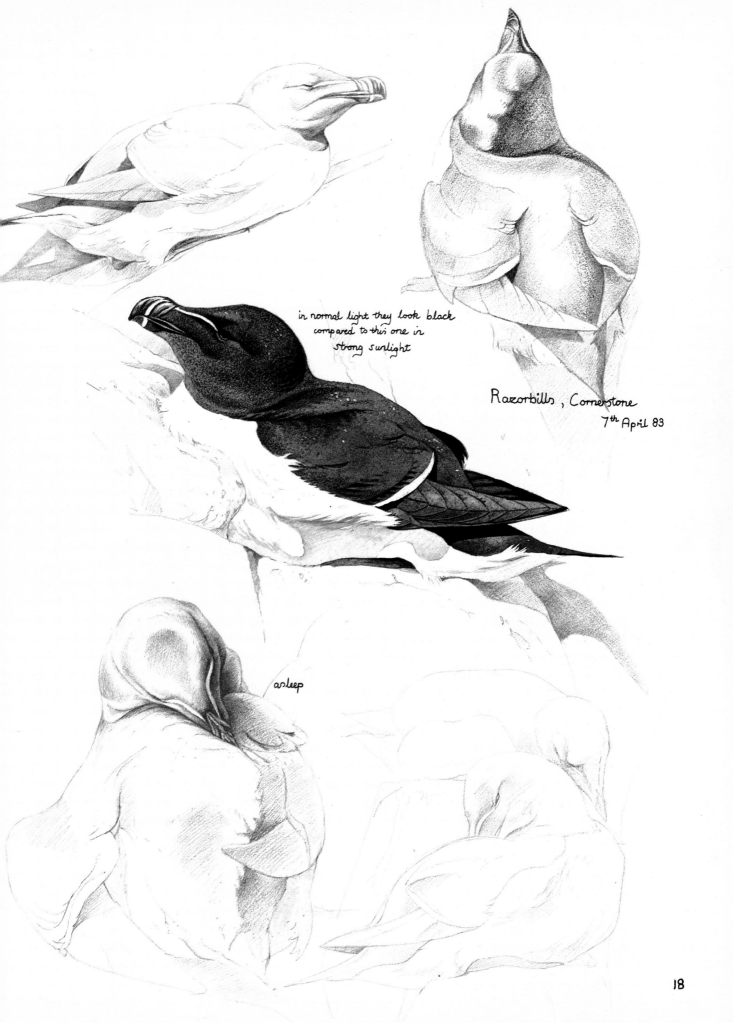

in normal light they look black
compared to this one in
strong sunlight

Razorbills, Cornerstone
7th April 83

asleep

Kittiwake

Razorbills at rest on ledges
Bishop's Cove   5th May 83

ruffling its crown.

very strong sunlight

Razorbills on ledges
by Mill Door 14th July 83

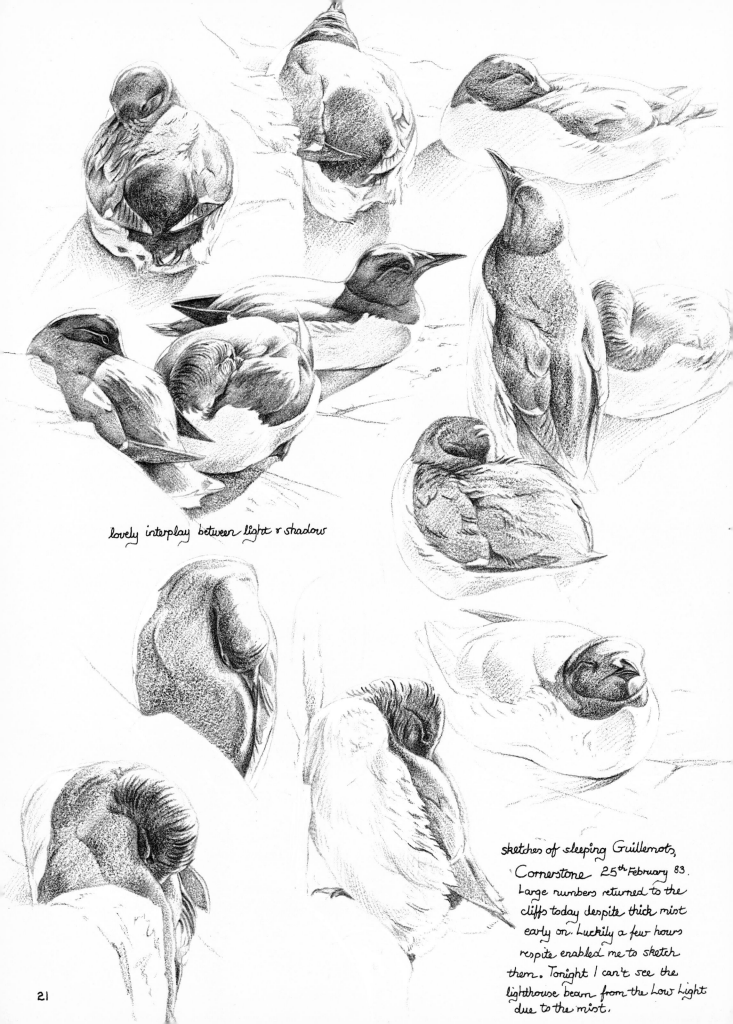

lovely interplay between light & shadow

sketches of sleeping Guillemots,
Cornerstone 25th February 83.
Large numbers returned to the
cliffs today despite thick mist
early on. Luckily a few hours
respite enabled me to sketch
them. Tonight I can't see the
lighthouse beam from the Low Light
due to the mist.

Most of the Guillemots are now
moulting into their summer plumage.

'bridled' form in
winter plumage.

Guillemots, Cornerstone 9th November 82

Sketched at dawn, the low yellow light strongly highlighting
the auks. Their feathers billowing in the gale force winds,
I was sketching them via a telescope from the relative
comfort of a hide. Periodically the guillemots return to
their nesting ledges from October onwards throughout
the winter

Puffin egg, laid underground in a burrow

all eggs life size

|———I———I 2 cm

Razorbill egg, 23rd May

a regurgitated Herring Gull
pellet consisting of shell and membrane
from a predated Guillemot egg

3 Guillemot eggs showing the wide
variation in markings and colouration.
'Pear shaped' so they roll in a circle
rather than off the flat cliff ledges.
(usually they are more soiled)

23

10th May 1983

Details of Guillemot chicks from 2 dead ones
15th & 16th June 83

@ 14 days old, a few
more days and it
would have left
for the sea with
the ♂ parent

all 1/1

pink gape

@ 9 days old

A force 8 westerly gale caused havoc with the
nesting seabirds on the lower ledges of the West Cliffs
during the 14th June. Between 4–500 Guillemot chicks/eggs
were washed away along with Shags and Kittiwakes.   24

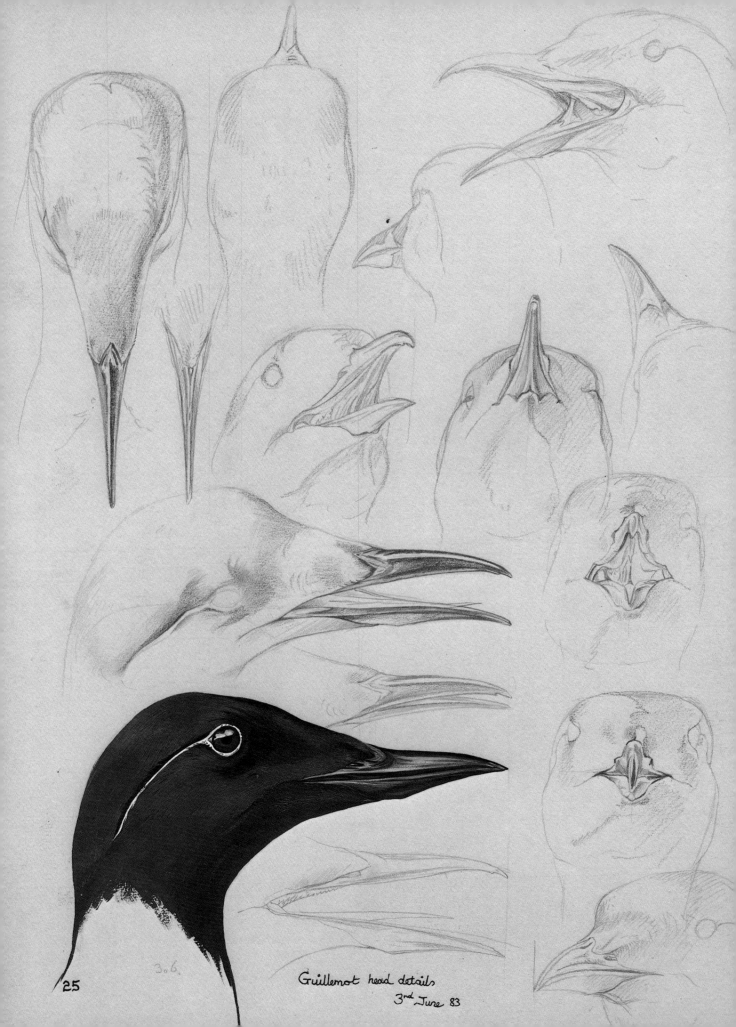

25

Guillemot head details
3rd June 83

back wet & covered
with droppings.
+ primaries tatty

S

H.H5

bluey brown
wet.

Guillemots, Dense
June 83

8430

F5

pool muddy water (ochre!)

H7

26

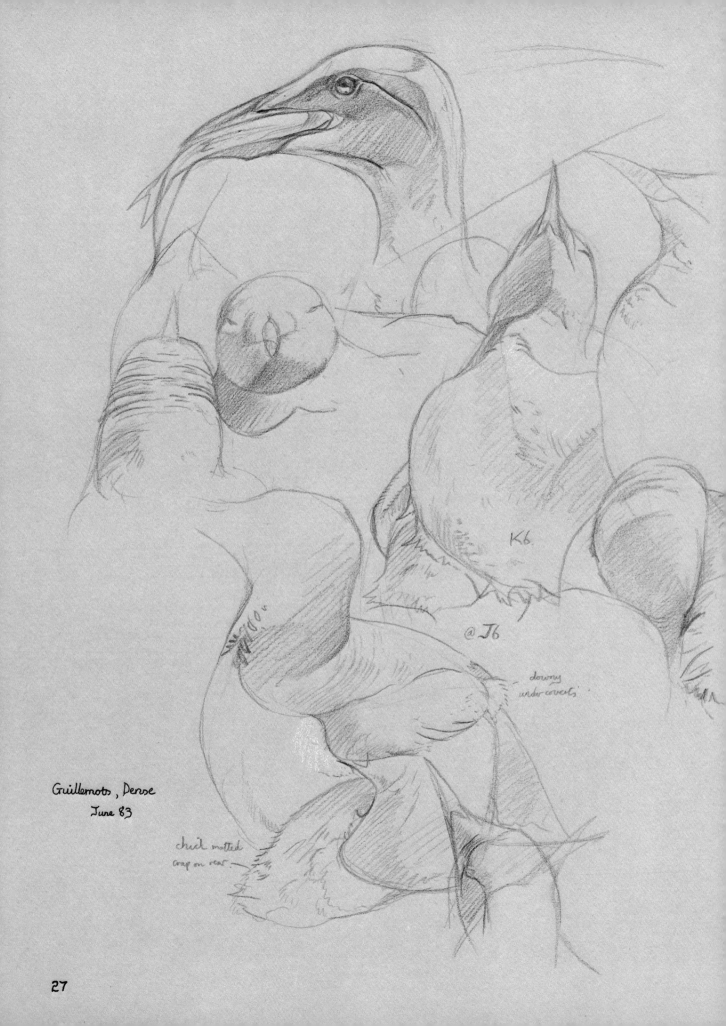

Guillemots, Dense
June 83

K6.

@ J6

downy
under coverts

chick matted
crap on rear

27

Guillemots, Derse - June 83

K7

28

29

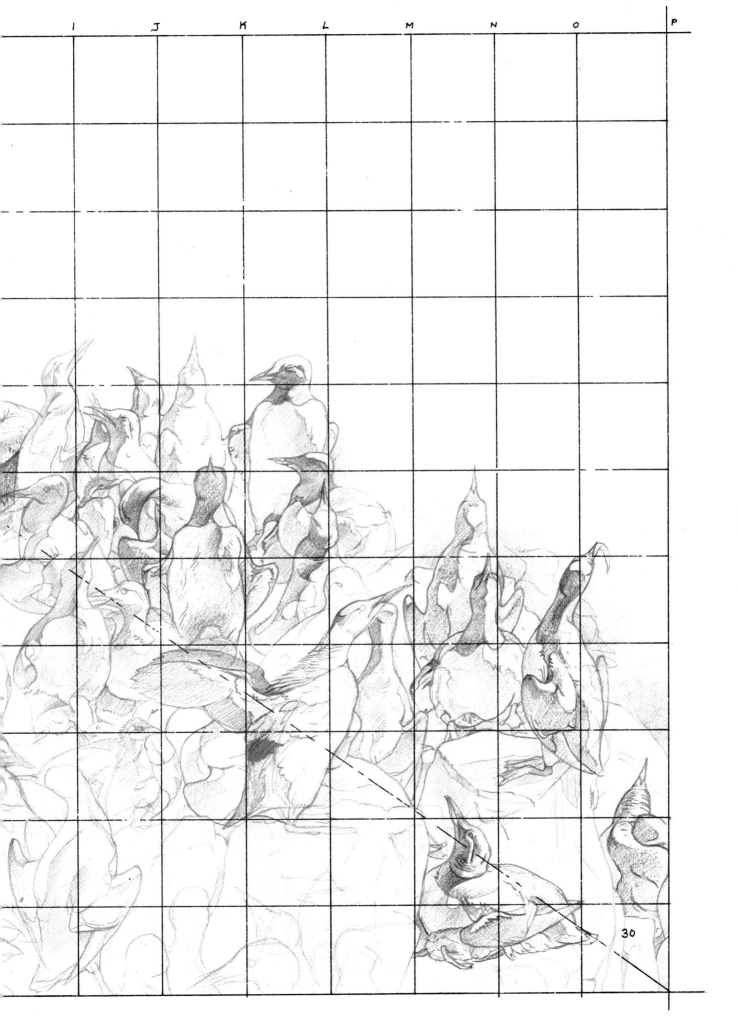

30

still some white winter
plumage feathers on throat
and neck, most are in full
summer plumage now.

oiled razorbill, found
dead in Pilgrims Haven
21ˢᵗ Dec 82

oiled guillemot which I
found whilst out 'dazzling' in
Kirkhaven,
drawn 2ⁿᵈ– 3ʳᵈ January 83

I had to put this bird down, frankly
it is the kindest way for oiled birds.
Usually death is caused by kidney
failure through ingestion of oil
by preening.

vainly attempting to wash
off the oil

getting waterlogged

preening in vain

heavily oiled

Found Dead

Guillemot	19
Razorbill	9
Puffin	1
Little Auk	3
Shag	2
Kittiwake	1
Herring Gull	2

A miserable group of oiled guillemots on the Maidens, 20th February 83. Heavy crude oil, often in large lumps was washed up all along the east shoreline, many more birds have small patches of oil on their plumage in addition to those found dead. These included Oystercatcher, Eider, Fulmar and Great Black-backed Gulls.

34

life size

3 newly hatched Eider ducklings, near the Low Light
5th July 83, I noticed that the ♀ was off the nest
so I sketched them quickly and took the hatched eggshell
and some down to paint later. They stayed in the
nest for at least 36 hours, the ♀ I presume waiting
for the 4th egg to hatch, till she led them down
to the sea. The unhatched egg contained a well
developed embryo. This nest was situated at the
entrance to a rabbit burrow and I often saw
rabbits running over the top of the incubating ♀.

♀ Eider, Kirkhaven June 83
This one has laid her eggs in a gull's nest.

36

♀ near the Low Light
6th May

♀ Eider nesting under the pipe (just by the Low Trap)
which carries compressed air
to the fog horns.

old rabbit snare

4th May 83

another ♀ on its clutch of eggs
right in the middle of a large
clump of sea campion, Kirkhaven.

lichen covered rocks

♀ Eider Duck on it's nest amongst the
rocks and sea campion, situated just
above the chapel

roosting oystercatcher (+ colour rings)

resting Eider ♂♂ in strong evening
sunlight accentuating their colours.

rock pale
grey

limpet

11th April 83

ochre
w. tip

barnacles.

lines of
mussels

seaweed   wash of
ochre +
more green/yellow
tip

+ white
highlight

♂ grazing mussels off the
reef

♂♂ displaying to a ♀
in the 'prone posture
15th April

39

adult with very worn tertials

Looking down on an Eider Duck and her
well grown brood, grazing off mussels at low
tide on the cliffs by Bishop Cove.
24th July 83

youngster wing stretching

40

♀

Immature ♂
asleep, Kirkhaven

26ᵗʰ January 83

Shags offshore on a windy day

white S
+ shaft

incubating birds

bluey-brown

white
tip to scap
+ tert

reflection of vegetation

panting in the sun.

Common Terns, North Plateau   30th May 83
drawn from a hide with the aid of a telescope.

young begging for food

preening

3rd summer
individual

Herring Gulls, Cobir Hole
17th August 83

43

Herring Gulls, Cobris Hole
18th August '83

panting in the strong sunlight
13ᵗʰ June '83

nest deep amongst
the thrift

Lesser Black-backed Gulls
Low Light

asleep, 2ⁿᵈ July

but always watchful

on guard duty

Lesser Black-backed Gulls
by Low Light, in strong
sunlight, 2nd July 83

preening

greater black-backed gull

adult, winter plumage.

30th December 82

adult herring gull, wp.
1/1

2nd yr kittiwake 1/1

drawn from a sick bird, 3rd March 83

47

this one below had
yellowish feet

First summer Kittiwakes
roosting and preening on E. Rona sheltering
from the force 6 westerlies, 2nd July 83.

Adult Great Black-backed Gull

48

looking down onto Kittiwake nests
on Bishop Stack, 24th July, 1983, the lower nest
contains 2 fledged chicks sleeping.

49

lovely shadows from
the fescue grass

Kittiwakes with chicks on the
cliffs in front of the Low Light
4th July 83

guarding their chicks against the
marauding Herring Gulls

50

Kittiwakes, drawn as they were resting on their nest ledges, 5ᵗʰ & 6ᵗʰ April 1983. They spent most of their time sleeping, occasionally waking up for bouts of frenzied display. The above birds were drawn in the shade compared to the lower birds with the lovely interplay of light and shadow.

dozing

wing stretching

Kittiwakes with large chicks
Greengates by Mill Door   July 83

begging for food

52

Fulmar pair, Cornerstone 5th March 83

cackling display with throat and
neck distended.

It took me 1½ hours patient observation to get
the basics for this sketch. The 'cackling display' occured frequently
during this period but usually for only 6 seconds at a time
with much head waving. Thus it took me a while to build up
an 'accurate' drawing.

sleeping Fulmars, ever watchful.

Fulmar portrait, near the Loose Tooth, 13th July,

a very tame individual sketched from 3-4 feet away.

54

the adults plumage at this time of year
is very worn and faded, shortly they will
moult most of their body feathers

Fulmars with their chicks, Cornerstone 7ᵗʰ Aug 83

55

Fulmar chick, below Low Light.
8th Aug, 83

mixed brown/green! more on
& blue black crown

ochre

settling down
on eggs.

♀ Lapwing, near the Beacon, 23rd May 83,
incubating a clutch of 4 eggs, sketched from a hide.

4 pairs failed to rear any young despite relaid clutches
due to the predations of crows and gulls.

Oystercatchers, a pair incubating 3 eggs by the Kettle Pools.
Sketched via a telescope from a hide @ 60 feet from the nest over
3 hours and 'worked up' later.

the pupil in the eye of an oystercatcher
never seems to be fully circular.

Danish Scurvy Grass

wing detail (upper) from a
dead Rock Pipit ¹⁄₁
Dec. 83

greater covert from a
normal rock pipit

grey(sh yellow)
brown
white tip
yellow/w/br
edging?
white tip

bright t/yel around eye
dull yellow-br
dull y-grey in front of eye
dark b. tip
orange + k.l.
pale y
yellow spot

prim - pale br + l.g edgings.

dull br/grey

sketch from bird in the hand
before releasing it, painted later
(24ᵗʰ Dec) from pencil colour notes.

lemon yellow

(edy
white y (out 3)
white y

reddy
brown

pinky / warm brown (highlights)
dull horn/b

yellowish soles.

leucistic Rock Pipit ¹⁄₁

October 1983

this aberrant individual (hatched this year) caused some
confusion amongst birdwatchers, normal birds have
dark olive-grey brown upperparts and streaks on the
underside.

nest lined with gull/kittiwake feathers

Brood of swallows in the Black Hut
by the Beacon, August 83. Nest on a beam
where plaster board had fallen off the ceiling
inside the hut. I sketched the chicks first then
the nest once the brood had flown.

60

2
OTHER WILDLIFE

Only six species of mammal were recorded on or around the Isle of May during 1983. The grey and common seals are dealt with in the fourth section of the book. Of the cetaceans only two species were recorded, the first a minke (lesser rorqual) whale sighted 300 yards off Pilgrims' Haven on 19 June by Mike Harris, which unfortunately I missed seeing, the second a common porpoise sadly washed up dead near the North Horn. This I sketched from several angles to illustrate the streamlined body that is beautifully adapted for life in the sea (61). The gulls later made short work of its carcase. The other two species, namely the rabbit and house mouse, are well established residents on the island. Rabbits have been present on the May from at least 1329, and at one time the warrens were the main source of revenue for the inhabitants of the island. Despite being ravaged by myxomatosis from time to time the rabbits are very abundant – probably over 2000 of them at their peak. They have to compete for burrows with the puffin population during the summer, and the activities of both species cause a fair bit of erosion. House mice were particularly abundant during 1983; Dr Graham Triggs, a research scientist on mice genetics, reckoned that there were between 4,000 to 6,000 mice present during the autumn. Larger than their mainland cousins they certainly made their presence felt despite the depredations of kestrels and short-eared owls.

My patience for them waned when, after a week's absence on the mainland, I returned to my room in the Low Light to find them everywhere. At least twenty erupted from a box of paper tissues which they had chewed to pieces to make a nest. They had also chewed my rubber, paint brushes and watercolour paper. Luckily I had taken all my drawings away the previous week or they, too, could have been part of the carnage. Action had to be taken before I was eaten out of house and home! I caught over forty mice in two days, in an attempt to stop them coming into the Low Light to spend the winter in relative comfort. Thankfully no rats have colonised the island, despite all the shipwrecks that have occurred, for they can cause much damage to nesting seabirds. A few rat carcases have been found but they are thought to have been brought over by gulls from rubbish dumps in Fife. A young stoat found dead was probably procured in similar circumstances. Bats, most likely pipistrelle, have been seen over the island only occasionally, and there is one record of a long-eared bat.

The beautiful small tortoiseshell is the most abundant butterfly on the island; it was particularly numerous during August 1983. On some hot days they seemed to be on every thistle head feeding on the nectar (66). They breed on the island and their numbers are reinforced by influxes of migrants. Many hibernate during the winter in the buildings on the island, including the Low Light Tower. The first I saw flying was on 1 April, a timely date for such a foolish venture on a cold, windy day! Some of the butterflies and moths, such as the Painted Lady which arrives from much further south, should technically be included in the migration chapter. Although they are more common some years than others, in 1983 they were few and far between. Those present in September were feeding on the remaining ragwort plants still in flower; at rest, their exquisite underwing, with its intricate pattern, is easily visible. The small number of large white and small white butterflies find the cabbages in the lighthouse keeper's vegetable plots much to their liking. A clouded yellow was recorded in 1982, the first ever sighted on the May. An angle shades moth fortuitously appeared by my arm whilst I was drawing by the light of a tilley lamp in the Low Light. I was able to paint it directly as it sat on the page, the cold making it a lethargic and perfectly motionless model (65). Many migrant moths are attracted to the artificial lights in the rooms.

At times, especially on wet days, the island seems to be literally crawling with the large common snail. Indeed you have to tread very carefully so as not to crunch any underfoot. The oystercatchers feed on these snails as do the song thrushes which occasionally breed here. Apparently no eels have been recorded in the Loch for a while but I saw at least six large eels there winding along just below the surface of the murky green water. Some appeared to be almost a yard in length.

The dry, hot summer months contributed to a plague of the berry bug or harvest mite from late August to October during 1983. The microscopic orange larvae burrow under the skin producing unbearably itchy spots. Usually, thank heavens, these pests are not so numerous, but most of the observatory visitors complained about them and I was particularly badly affected. I was literally a sitting target for them as I was often stationary for a long while sketching in one place. At one time I had over fifty bites on one of my forearms. Even the rabbits had problems with them, many having orange-tipped ears because they were so infested.

Island life is most prolific below the high-water mark down to the depths of the sea. I have chosen here only a few examples that are of interest to me. Most of these sea creatures are terribly complicated and too microscopic to draw; they would drive me to distraction. The lobster, however, was a fascinating subject, one I had never drawn before, and it was a delight to discover all its intricate forms (63). This one was missing two legs and most of one antennae. Fishermen from Anstruther set lines of creels (or pots) off the island to catch lobsters and edible crabs (partins). The shore crabs are most evident at night in the deep rock pools and as they forage amongst the seaweed. The rock pools are also home for many fish such as the shanny, probably the most common. Usually all one gets is a fleeting glimpse as the fish dart from the shallow edge down to the depths of the pool when disturbed. Shells are really limited to the two sandy beaches at Kirkhaven and Silver Sands. I particularly enjoy drawing broken and worn shells that reveal the spiral structure inside, as well as those encrusted with barnacles and tube-worm cases. I collected the sea urchins on a spring tide at low water along the foot of the cliffs by Horse Hole (77). This tidal state is the best time to explore the seashore on foot for it allows access to areas that are normally underwater. As one slips and slides amongst all the various seaweeds one finds creatures ranging from starfish to sponges. The sea urchins are particularly beautiful when all the tube feet appear from amongst the spines and wave in the water.

The flora of the island have been well documented for a hundred years or more. The vegetation has been much modified by gull colonies, and areas such as Rona, once covered in extensive growth of thrift, are now carpeted with chickweed, sorrel, sea campion and scentless mayweed. One of the first plants to flower is the lesser celandine which is especially abundant in the Bield area in front of the Low Light. Its pretty bright yellow flowers are a joy to see in the early spring (68). The thrift is the most beautiful of all the May's flowers. Their cushion-like bases of narrow leaves are adorned with a mass of purple-pink flowers on leafless stalks (69). Although not a patch on their former glory they still cover some areas of the island. The most prominent flowering plant is the sea campion (70). On the South Plateau and the northern slopes of the Loch they form a shimmering sea of white flowers in midsummer. Cracks amongst the rocks provide an anchorage for the English stonecrop, a plant with succulent leaves and white star-shaped flowers (73). The silverweed is a common plant that can be found right down to the high-water mark. It sends out runners from the main root which support the yellow flowers (74). It is so named because of the silvery coloured underside to the leaves. The henbane, a highly poisonous plant, is rapidly colonising the coal ash area around the Beacon and behind the Low Light. They have stout, hairy, sticky stems up to three feet high with a row of yellow flowers marked with a vein-like network of purple (71). The walls of the ruined chapel are host to many plants such as the sea spleen-wort fern which I painted there: it also grows in ceiling cracks in the cliff caves (73). The walls of the Chapel are bedecked with thrift, dandelion and scurvy grass. The aptly named thistle field is thick with thistles and wood burdock in the autumn, their purple flowers providing a great source of food for the numerous small tortoiseshell butterflies. Ragwort is the last flower to remain in bloom until it too is 'burnt off' by the salt spray from the first winter gales.

The tenacity of some plants in this stormy maritime climate is extraordinary. Despite all the gales and salt spray, a specimen of cultivated fig has been growing out of a small rock fissure on the edge of the Burrian, a marked contrast to its usual habitat in warmer climes. It must have grown from a seed dropped by a bird. Sea wormwood, a plant close to its northern limit on the May, grows in one small area on the sheer cliffs by the South Horn and survives an almost constant buffeting (72). Two specimens of the northern marsh orchid were growing right by the Holyman's Road path in 1983 and one managed to survive being trampled underfoot to flower briefly. Lovage is normally confined to the south slopes of the Loch because of its intolerance of salt spray but in 1983 a single plant managed to flower halfway down the cliffs by the Maiden's Bed.

There are only a few trees on the island, brought over to provide cover for the migrant birds, and also to attract them into the Heligoland traps. The trees consist of sycamore, sitka spruce, willow and elder which can only survive behind the shelter of walls. Hops have also been introduced and they provide splendid cover as they climb up the wooden posts supporting the traps.

During the stark winter months the dour greys and browns of the rocks and vegetation are somewhat relieved by the yellow and green lichens with which they become encrusted. Some of the rocks, especially around the jetty at Kirkhaven, are really beautiful, coated with large growths of ramalina and xanthoria parientina. I painted a few of the more obvious species, marvelling at their complex structure yet not realising how difficult their subsequent identification would prove.

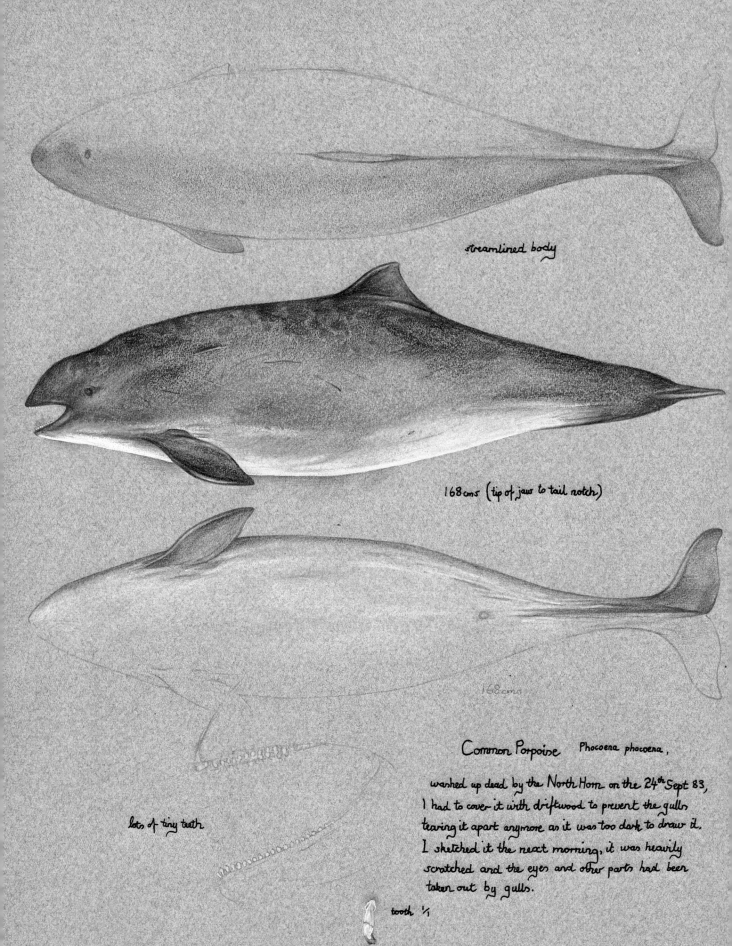

streamlined body

168 cms (tip of jaw to tail notch)

168 cms

lots of tiny teeth

Common Porpoise   *Phocoena phocoena*,

washed up dead by the North Horn on the 24th Sept 83,
I had to cover it with driftwood to prevent the gulls
tearing it apart anymore as it was too dark to draw it.
I sketched it the next morning, it was heavily
scratched and the eyes and other parts had been
taken out by gulls.

tooth ¹/₁

hiding behind a rock

Shanny, c ½, Blennius pholis (nick-named Fanny), they are very
common in the rock pools on the May
.

pelvic fins

31st Dec 83, some fish from a salt
water aquarium belonging to the Marquiss family

Hermit Crab.

Sea Scorpion

Common Lobster @²⁄₃ life size
15ᵗʰ August, 83.

caught in a creel off the island and released after
I had painted it as it was below the
commercial size. This one had been
damaged on its left side with 2 legs
and most of the antennae missing. It
should be able to replace these though.

Shore Crab (life size) C. maenas
from Kirkhaven   28th Nov, 83

Angle Shades Moth, *Phlogophora meticulosa* ¹/₁
26th October 1983

This fortuitously appeared at the Low Light window today, it
is apparently the first Isle of May record since 1914!

Small Tortoiseshell
feeding on a Spear Thistle Cirsium vulgare

12th August, 83

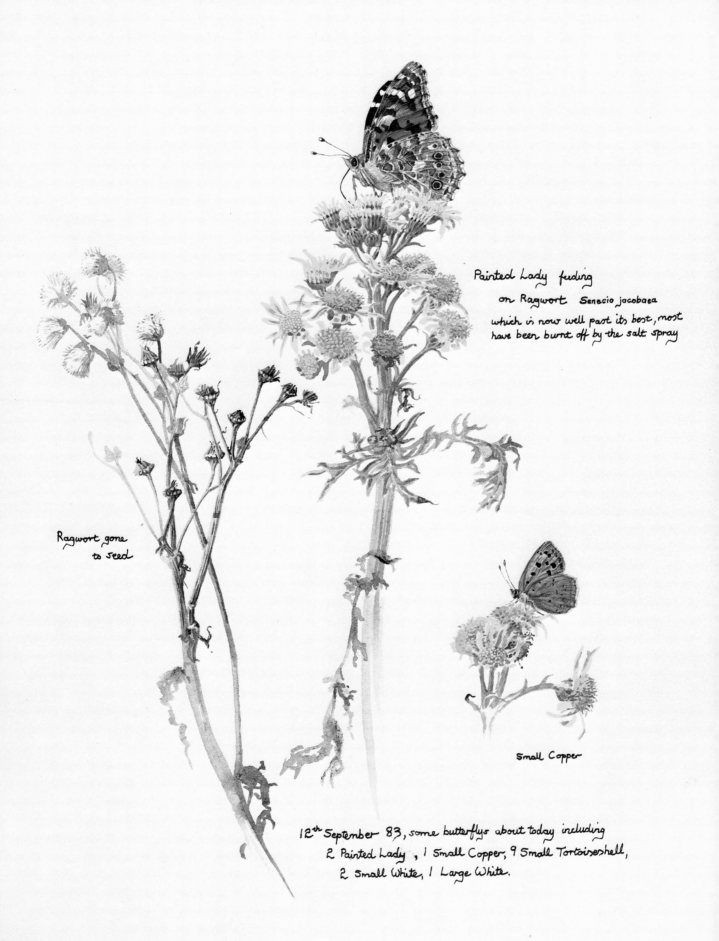

Painted Lady feeding
on Ragwort Senecio jacobaea
which is now well past its best, most
have been burnt off by the salt spray

Ragwort gone
to seed

Small Copper

12ᵗʰ September 83, some butterflys about today including
    2 Painted Lady , 1 Small Copper, 9 Small Tortoiseshell,
    2 Small White, 1 Large White.

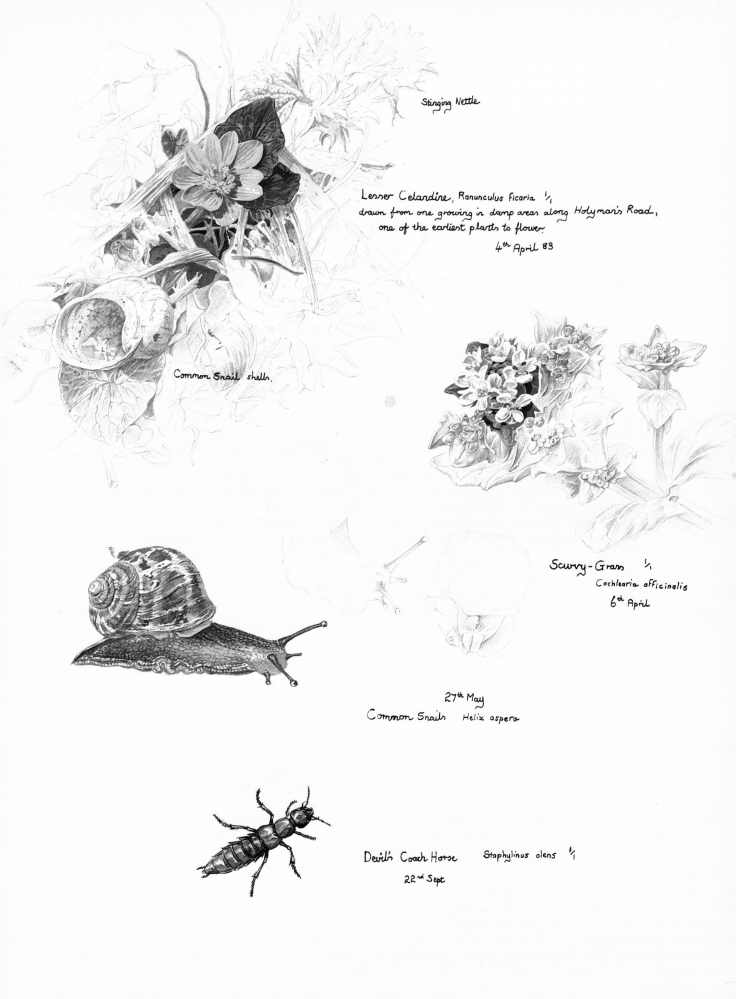

Stinging Nettle

Lesser Celandine, Ranunculus Ficaria $\frac{1}{1}$
drawn from one growing in damp areas along Holyman's Road,
one of the earliest plants to flower.
4ᵗʰ April 83

Common Snail shells.

Scurvy-Grass $\frac{1}{1}$
Cochlearia officinalis
6ᵗʰ April

27ᵗʰ May
Common Snails   Helix aspera

Devil's Coach Horse   Staphylinus olens $\frac{1}{1}$
22ⁿᵈ Sept

Thrift *Armeria maritima,*
12th July 83

69

Henbane Hyoscyamus niger
9th Aug, 83

seed pods of Henbane, 19th Sept.

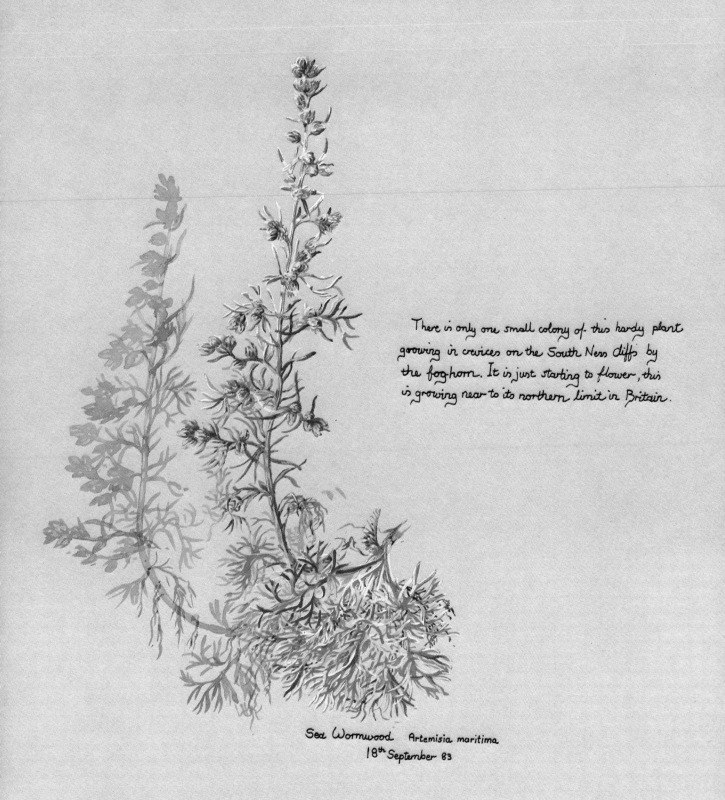

There is only one small colony of this hardy plant
growing in crevices on the South Ness cliffs by
the foghorn. It is just starting to flower, this
is growing near to its northern limit in Britain.

Sea Wormwood   Artemisia maritima
18th September 83

English Stonecrop, $\frac{1}{1}$ Sedum anglicum

Holyman's Road. 12ᵗʰ August

Sea Spleenwort Fern $\frac{1}{1}$ Asplenium marinum

growing on the walls of the chapel,

14ᵗʰ August 83

Silverweed   Potentilla anserina

Kirkhaven, 10ᵗʰ August 83.

A creeping plant with long runners growing on the upper beach and elsewhere.

74

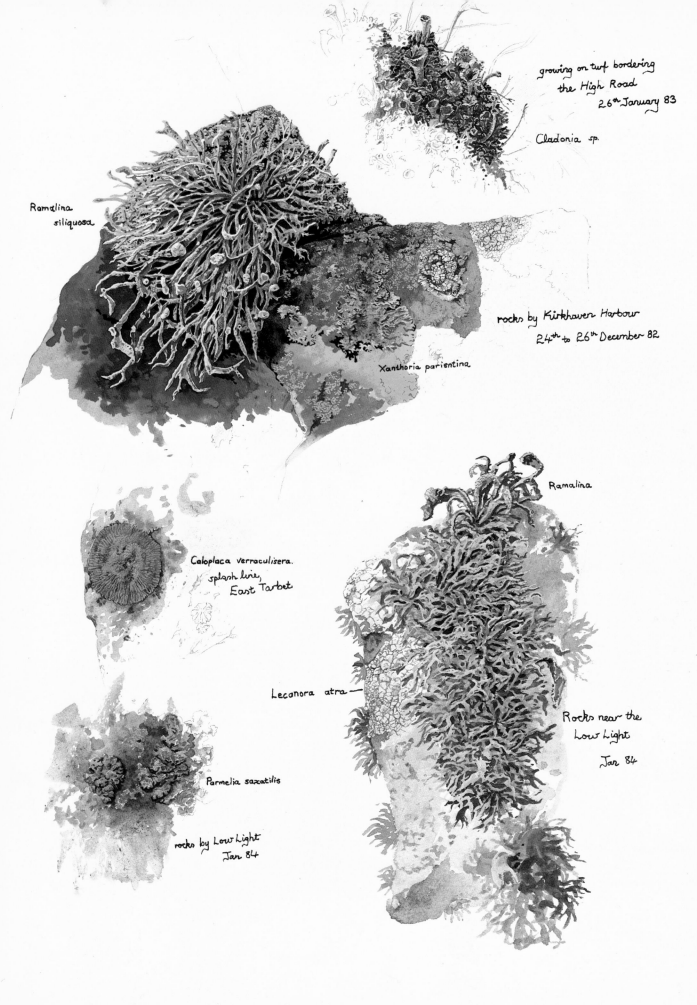

growing on turf bordering
the High Road
26ᵗʰ January 83

Cladonia sp.

Ramalina
siliquosa

rocks by Kirkhaven Harbour
24ᵗʰ to 26ᵗʰ December 82

Xanthoria parientina

Ramalina

Caloplaca verraculisera.
splash line,
East Tarbet

Lecanora atra

Rocks near the
Low Light
Jan 84

Parmelia saxatilis

rocks by Low Light
Jan 84

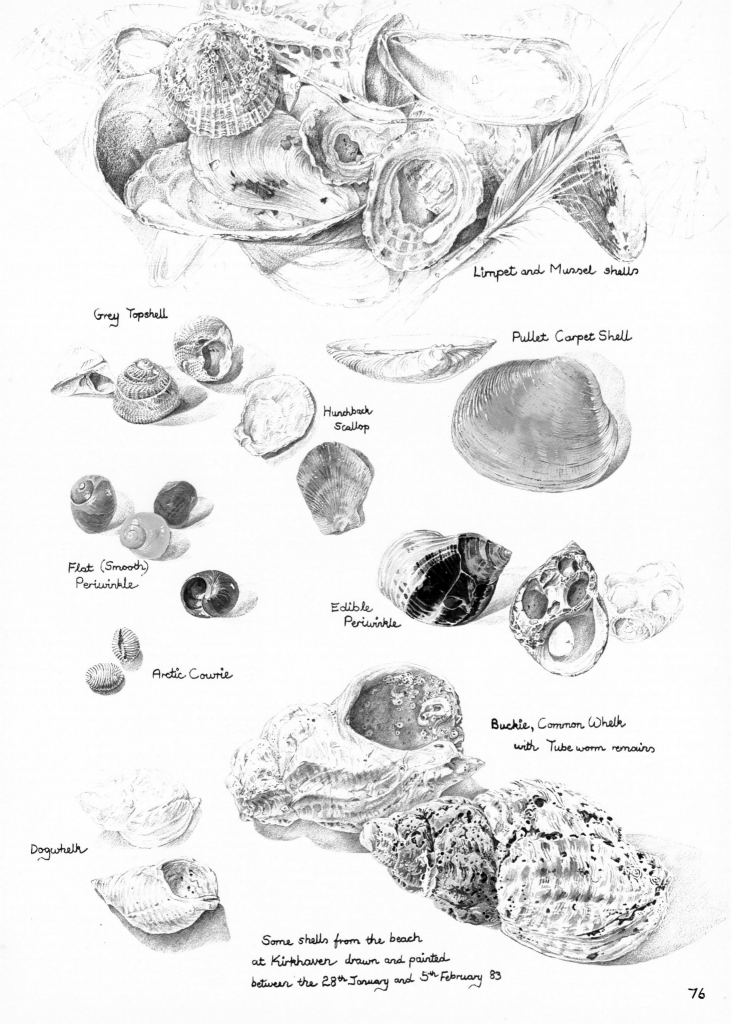

Limpet and Mussel shells

Grey Topshell

Pullet Carpet Shell

Hunchback
Scallop

Flat (Smooth)
Periwinkle

Edible
Periwinkle

Arctic Cowrie

Buckie, Common Whelk
with Tube worm remains

Dogwhelk

Some shells from the beach
at Kirkhaven drawn and painted
between the 28th January and 5th February 83

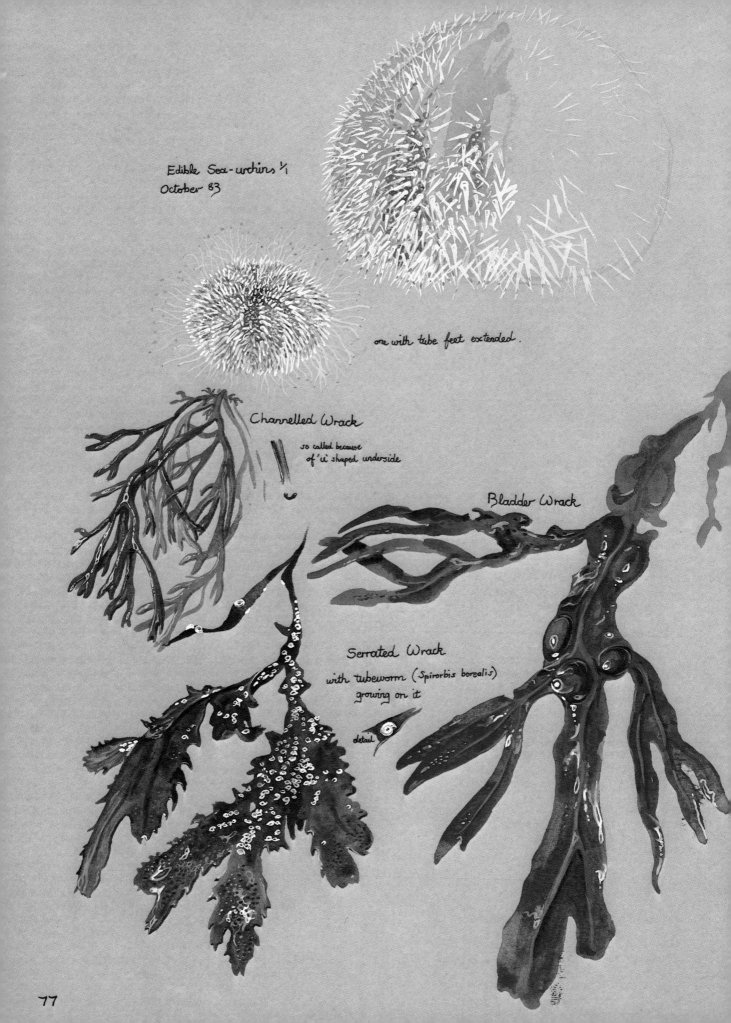

Edible Sea-urchins ½
October 83

one with tube feet extended.

Channelled Wrack

so called because
of 'u' shaped underside

Bladder Wrack

Serrated Wrack

with tubeworm (Spirorbis borealis)
growing on it

detail

77

House Mice   Mus musculus
September 1984
large island form, drawn whilst
captive for a short period inside a
clear plastic box. All around life size

a large individual, pelage detail.

grooming

78

Sunning & grooming rabbits by the Low Light.

31st August 83

Rabbit head detail (life size)
1st September 83

3
MIGRATION

Migration is one of the great spectacles of the bird world. As an example of what I mean, try and picture the sight of thousands of migrant birds carpeting the tiny island of May, seeking food or shelter in almost every nook and cranny, on every stem of thistle and dock, while overhead thousands more birds move continuously past from dawn till dusk, their calls scarcely audible above the tumult of wind and rain. That was the scene on the island between 11 and 13 October 1982 after an area of high pressure over north Russia and Scandinavia, and a depression over the North Sea with a strong easterly airstream, had forced many migrants off course. Having started out on their journey in good weather, they had become disoriented by the overcast skies near the Scottish coastline and had taken refuge on the May. It was one of the most spectacular falls of migrants ever recorded on the island. On 11 October alone, over 25,000 birds were present, including a staggering 15,000 goldcrest, 4,000 robin, 4,000 fieldfare, 800 redwing, 600 blackcap, 400 brambling and 200 chiff-chaff. Also present during this three-day period were yellow-browed warbler, three Pallas's warbler, barred warbler, bluethroats, shore lark, three great grey shrikes and a rough-legged buzzard. Some robins that were caught bore Finnish and Norwegian rings, giving an indication of the origin of these migrants. Tragically, the driving wind and rain proved too much for many of the goldcrests. Exhausted on arrival, they succumbed soon afterwards. Many more must have perished at sea before finding the May. Indeed it never ceases to amaze me how such a tiny bird, weighing only between five and six grams, can cross the North Sea in such adverse conditions. The large movement of birds that October was recorded from as far north as the Shetlands down to Norfolk, with the main concentrations being in south-east Scotland and north-east England. The total numbers involved must have been vast.

Unfortunately there were no migrant falls on this scale during 1983, for the predominantly westerly winds during September and October held back much of the usual migration. There was quite a good fall on 29 September (details opposite) but it was on a smaller scale to the previous year. Most visitors to the Observatory hope they will coincide with a large

Observatory: Isle of May   **Date:** 29th September, 1983

Observers: Keith Brockie, John Callion, Wendy Mattingley, Dennis White, Chris Campbell, Owen Hayward, Stuart Green.

WEATHER: wind direction and strength; visibility; precipitation (incl. fog); cloud amount, etc. Give time of changes in G.M.T.

NE by E force 4-5, Vis = Poor to Moderate, Intermittant Showers (heavy at times), 8/8 Cloud Cover

RINGING: species totals; numbers weighed/measured (in brackets); re traps, with dates of first capture.

Purple Sandpiper 11 (inc 2 ret), Turnstone 19 (inc 1 cont, 1 ret), Redshank 7 (inc 1 ret), Golden Plover 1, Curlew 2 Eider 4, GBb Gull 1, Goldcrest 13, Willow Warbler 1, Chiffchaff 2, Blackcap 17, Robin 45, Garden Warbler 4, Redstart 10, Siskin 2, Reed Bunting 2, Lesser Whitethroat 1, Blackbird 1, Ring Ouzel 1 Song Thrush 20, Redwing 1 (165)

MOVEMENTS: significant arrivals/departures; visible movements with times G.M.T. of arrivals/departures; flight directions; lighthouse observations; sea-passage.

Arrivals

1 Shellduck	1 Whinchat	15 Spotted Flycatcher
3 Wigeon	24 Ring Ouzel	1 Pied Flycatcher
1 Peregrine (adult)	30 Blackbird	10 Chaffinch
1 Water Rail	55 Fieldfare	50 Brambling
2 Golden Plover	1000 Song Thrush (inc pass)	12 Siskin
5 Lapwing	600 Redwing	4 Twite
4 Snipe	1 Reed Warbler	2 Scarlet Rosefinch
2 Wood Pigeon	1 Barred Warbler	10 Reed Bunting
c 60 Skylark	6 Lesser Whitethroat	
6 Tree Pipit	3 Common Whitethroat	Passage
c 250 Meadow Pipit	100+ Garden Warbler	1 Red-throated Diver S→
1 Dunnock	250+ Blackcap	1 Sooty Shearwater N→
600+ Robin	10 Chiffchaff	11 Manx Shearwater N→
1 Black Redstart	40 Willow Warbler	4 Arctic Skua
100 Redstart	3000 Goldcrest	

LEPIDOPTERA MIGRATION: approx. numbers, flight directions.

1 Small Tortoiseshell

The migration log page for the 29th September detailing the bird movements on the island. This was the best day in terms of numbers of migrants recorded during 1983. The next few days brought further birds including wryneck, red-breasted flycatcher, shore larks and a large passage of barnacle geese

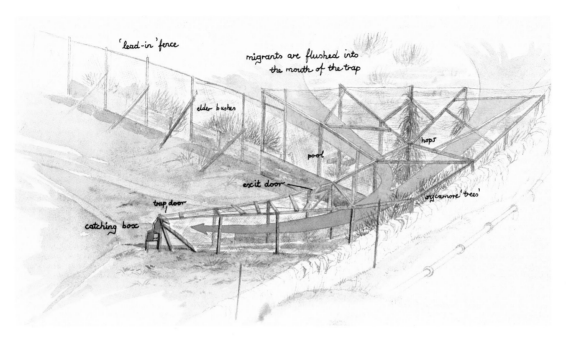

Diagram of a Heligoland Trap, namely the Low Trap, in April. The birds are attracted to the trap area by the bushes and freshwater pool. They are periodically flushed into the mouth of the trap by approaching bird ringers and pushed into the catching box at the end of the funnel. This has a sloping glass frontage which the birds think they can escape through. The catching box trap-door is then shut and the birds extracted for ringing and subsequent release. Most of the migrants ringed are caught in this manner

```
RECORD OF RINGED BIRD                                                      Ring number
                              STAVANGER                       NOS   7170802
Species:
Art, soor BLACKBIRD                                                 CODES 18705

Age/Sex 2ND YEAR           FEMALE              5 CODES CODES  CODES  CODES  CODES
Alter/Kon

Ringing Information:        Date accuracy (time)                    Co-ordinate accuracy
Beringt
Bagué       31  MAR  81    ±               59  4'N   10 32'E    ±
Geringd
Anillada
Marxet     STORE FAERDER,
Rengästettu
Merket     TJOME,                                                  CODES  CODES
- - - -VEST-AGDER, NORWAY - - - - - - - - - - - - - - - - - - N036 - -
Finding Information:        Date accuracy (time)                    Co-ordinate accuracy
Gefunden
Repris      5  NOV  82     ±               56 11'N    2 33'W    ±
Gevonden
Recuperada
Fundet     ISLE OF MAY,
Löytö
Funnet                                                            CODES  CODES
           FIFE REGION, SCOTLAND          Finding Details:  Remarques  Bemærkning GBFR
                                                           Bijzonderheden Lisätietoja
                                                        1  Detalles   Bemerkungen
           ISLE OF MAY B.O. ✓             8 CODES CODES CODES B CODES CODES CODES
                                          CONTROLLED
                                          4F/ 134 MM/ 102.5 G/ 1100.

                                          Distance:    Direction:    Duration:
                                          841 KM       248 DEG       584 DAYS
Ringer:
Finder's copy of details recorded in our permanent file. If you find any factual    CODES  CODES  CODES
error please inform us, B.T.O. Beech Grove, Tring, Herts HP23 5NR, England.          270383 RHODES  CODES
```

Example of a ringing recovery print-out which the finder of the ringed bird will receive upon sending in the relevant details to the address inscribed on the ring (in this case Stavanger Museum, Norway)

fall of migrants, with plenty of rare species turning up, and there is always an air of excitement on hearing the radio shipping forecast, when the weather patterns, always unpredictable, appear conducive to a fall of migrants. Undoubtedly the mainland coast of Scotland attracts many migrants but these quickly disperse into the extensive fields and woodland and are much more difficult to locate. The small size of the May and its lack of cover mean that many of the migrants concentrate around the trapping areas and are thus relatively easy to spot – although in strong east winds and rain many migrants, such as goldcrests, seek shelter on the West Cliffs, and the air becomes full of their high-pitched calls sounding above the thunderous seas below.

The main daily function of those staying in the Bird Observatory during the spring and autumn is to document and ring (only if a qualified ringer is present) the migrants on the island. The migration log and daily census book are filled in with weather conditions, numbers ringed, arrivals, departures and passage migrants, i.e. birds flying offshore or over the island. Up until December 1983 some 255 species of bird had been recorded on the island including such diverse species as great shearwater, red-footed falcon, Pallas's sandgrouse, Siberian thrush, black-eared wheatear, Sardinian warbler, citrine wagtail, woodchat shrike, pine grosbeak and yellow-breasted bunting, to name but a few. During 1983 some 159 species were recorded including three new species for the island – honey buzzard, grey phalarope and rose-coloured starling. I added three new species to my own list – the rose-coloured starling, firecrest and marsh warbler. Much to my chagrin, during one week in May when I was on the mainland I missed three species which I had never seen before – the common and thrush nightingale and a woodlark. (I have included a page of ringing recoveries to illustrate some of these cosmopolitan travellers.) Some other species of interest not illustrated or mentioned in the following pages include long-tailed and Pomarine skuas, Brent geese (dark-bellied race), quail, Iceland gull and a rustic bunting. I have, at times, also had tantalizing glimpses of a female or immature male red-flanked bluetail.

Some 8,805 birds were ringed during 1983, the majority of them seabird pulli (chicks). Heligoland traps, named after the island off West Germany where the trap was first devised, catch most of the migrants. The trap basically consists of a wire mesh-covered framework with a funnel-shaped opening which narrows progressively to a catching box with a hinged flap which is dropped when all the birds enter the box. The birds are attracted to the bushes or the other cover at the mouth of the trap and are driven into the trap up the funnel. The birds caught in this way are taken out of the box and back to the ringing hut where they are processed. A bird is first identified and ringed with the appropriate sized ring, next it is aged and sexed according to *Svensson* (a field guide for birds in the hand), measured –

Selected ringing recoveries illustrating origins and destinations of migrating birds

Species	Ringing Location	Where Recovered
Kittiwake	Isle of May, 8.7.74 as pullus.	Fogo, Newfoundland, Canada, 20.9.76 (shot).
Great Black-backed Gull	Isle of May, 16.12.74 as adult.	More og Romsdal, Norway, 5.6.81 (shot).
Long-eared Owl	Isle of May, 20.10.79.	Spier, Drente, Netherlands, 3.2.82 (road casualty).
Meadow Pipit	Isle of May, 23.9.80 as adult.	Larache, Morocco, 3.2.81 (found dead).
Robin	Turku-Pori, Finland, 18.9.82 as juvenile.	Isle of May, 11.10.82 (controlled).
Bluethroat	Isle of May, 17.5.77 as 2nd Yr ♂.	Burs, Gotland, Sweden, 27.9.78 (found dead).
Redstart	Isle of May, 25.9.65 as 1st Yr.	Zanzur, Nr Tripoli, Libya, 6.4.66 (trapped).
Wheatear	Isle of May, 6.9.82 as 1st Yr.	Bechar, Algeria, 9.4.83 (trapped).
Blackbird	Heligoland, F.R. Germany 9.3.81.	Isle of May, 5.11.82 (controlled).
Song Thrush	Isle of May, 30.4.78.	Bologna, Italy, 10.12.79 (shot).
Willow Warbler	Epse, Gelderland, Netherlands, 24.4.80 as adult.	Isle of May, 3.5.80 (controlled).
Goldcrest	Kallskar, Finland, 5.10.75 as 1st Yr ♂.	Isle of May, 22.10.75 (controlled).
Red-breasted Flycatcher	Lågskär, Finland, 3.9.75 as 1st Yr.	Isle of May, 10.10.75 (controlled).
Pied Flycatcher	Viljandi, Estoniya, U.S.S.R., 27.6.80 as pullus ♂.	Isle of May, 13.5.81 (controlled).

(controlled means caught and released)

usually just the wing length – weighed and released back to the wild. The value of ringing is very important. It helps us to find out about where and when the birds were born, their migration routes, wintering quarters etc. Linking this with information from bird observatories elsewhere in the country one can form some sort of migratory pattern. Alas, however, the recovery rate of birds, especially small passerines, is very low. Some of those that we do recover are species of small birds that have had to run the gauntlet of 'sportsmen' in the Mediterranean who shoot or trap them in their hundreds.

In contrast to the large falls of wayward migrants, normal migration, termed 'coasting', occurs when birds arrive in the spring and depart in the autumn. This is especially evident in August as birds move south along the coast of Scotland. Flocks of meadow pipits and a constant trickle of willow warblers enter the traps on the island. The juvenile cuckoo which I drew was just such a migrant, on its way gradually south to its wintering grounds in Africa. The thrushes which arrive in large numbers are emigrants from Scandinavia coming to Britain to overwinter. Generally most pass over during the night, their calls the only indication of their presence. Only during adverse weather conditions, when they become tired and lost, do they alight on the island, which is their first available landfall. Most stay only a short while to rest and refuel before moving on. During November 1982 thousands of blackbirds moved through and three of us ringed over 500 of them in two days. Fog is not generally good for our purposes as most migrants then tend to miss the island and land on the nearby mainland – as was the case for many days in the spring of 1983 at Fife Ness.

A bird in the hand gives me a great opportunity to draw close up – especially to get detail drawings of salient features of some of the rarer species. One of the most amazing birds in the hand is the wryneck with its fantastic dead-leaf plumage pattern. The snake-like contortions it makes with its head and neck whilst raising its crown feathers are extraordinary (91). With its hooked and notched bill the vicious great grey shrike can draw blood upon any lapse of concentration (83). The large fiery eyes of the long-eared owl seem enraged at being caught, and it clicks its bill and raises its ear-tufts in defiance (95–6). The rich metallic blue gorget of a male red-spotted bluethroat is one of the most beautiful of the Scandinavian migrants (85). The detail of a Pallas's warbler is similar to the goldcrest and firecrest, but it is a rare gem from southern Siberia and, on the May, is thousands of miles off course from its wintering quarters (90). The irridescent plumage of a lapwing makes me realize just how colourful the young of this species are (100).

Most birds will relax in the hand if held motionless for a short period free from outside disturbance. I quickly draw and measure the bird (with dividers), using pencil in most cases, with relevant colour notes. These I paint up later while the bird is still fresh in my mind. The bird is then released unharmed (which couldn't be said for me after handling the owl and shrike!). Birds found freshly dead such as the gannet and bonxie are put to good use with plumage maps depicting, say, the proportions and feather structure of an outstretched wing (108–112).

Another method of catching coastal birds is by dazzling them. This involves using a quartz halogen handlamp connected to a motor-cycle battery carried inside an acid-proof container in a rucksack. The ideal conditions are a pitch-black night with no moon, preferably with rain and some wind to disguise any noise. Just trying to keep one's feet on the slippery rocks and seaweed and avoiding the occasional rogue wave is a challenge. Sometimes one is startled by 'rocks' which come to life at one's approach – seals disturbed from their slumber that charge wildly into the sea, sets of their curious companions' eyes reflecting the torch-light like stars in the darkness.

The waders feed along the water's edge at low tide on the reefs, even at night, with some roosting on the pools above. I try and transfix them in the beam of the lamp and creep forward hoping to catch them in the landing net. It's not as easy as it sounds and many fly off before I can get to them. During 1983, however, I did catch nearly 300 waders by this method, mostly purple sandpiper and turnstone, but also, amongst others, two whimbrel, golden plover and teal.

A brood of Purple Sandpipers only one day old with colour rings

Hardangervidda, 1st week of June 1982.
Purple Sandpipers are present in these conditions

I have a special interest in the purple sandpiper (101), a small wader which feeds amongst the reefs. It is present every month except June in numbers varying from around 50 to an autumn maximum of 400. One had previously been recovered on the Hardangervidda Plateau in southern Norway, an area of tundra above 3000 feet. In order to prove more conclusively that this area was the breeding ground for some of the purple sandpipers on the east coast of Scotland I took part in expeditions to the Hardangervidda in 1978, 1980 and 1982. Here we caught adults and their chicks, individually marking them with combinations of three colour rings plus the usual metal ring. The contrast between the coastal wintering areas and the high tundra breeding grounds, often still covered in snow when they arrive, is very marked. To date, three of these birds have been sighted on the Isle of May, two of them in consecutive years. In addition, others have been observed down the east coast of Britain from Aberdeen to Scarborough.

Of great interest was a purple sandpiper I caught in August 1983 with a ring which was so badly worn that no digits were discernible. I sent the ring off and it was etched by chemicals to reveal its number. It had been ringed on the island in September 1969 as a moulting adult, thus making it at least fifteen years old – a British longevity record. Another I caught in May had been ringed in November 1977 on the Isle of Man. All this knowledge helps me to form a more complete picture of a bird's lifestyle. The Observatory's ringing recoveries certainly prove that the Isle of May's birds are far from insular.

Red Backed Shrike   Juv ♂ ³⁄₂ plumage
Lanius collurio   detail
wg 95mm   only.
t 25mm   Isle of May 16.8.77
b 18mm   drawn from a bird in the
wt 27.4   hand.

outer tail feather.

Greenish Warbler ⁷⁄₄ Isle of May 14.8
Phylloscopus trochiloides viridanus

wing 59.5mm   4ᵗʰ longest
bill 12mm   2ⁿᵈ - 6.5
tarsus 19mm   3ʳᵈ - 0.5
wt 6.6grms   5ᵗʰ - 2.5
                 6ᵗʰ - 5.5

tarsus darkish
grey brown horn

wing formula ⁷⁄₁

emarginated 13456.

Some early sketches drawn on the
Isle of May in August 1977.
Above, a juvenile ♂ Red-backed
Shrike x ³⁄₂, 1983 was a poor year
with only one (♀) seen in early
June during a spell of easterly
winds.
Opposite, a Greenish Warbler x ⁷⁄₄,
a rare vagrant to Britain which I
caught on the 14ᵗʰ of August.

81

Grey Phalarope. ½ 1st year   8.9.1977
intermediate plumage.

wing 129 mm   weight 59.8
bill 23.5 mm
tar 25 mm
tail 63 mm

— bill broadens then
goes to a triangular point

(brown edged black feathers
extend up the nape   moulting into winter...)

some yellow
at base

dusky tint
in rest ...

partial webbed feet

inside

buoyant swimmer

fed near floating seaweed
constantly picking at water
& weed, twirling round as
it did so

A Grey Phalarope was spotted off the South Ness by Dave Pullan
on the 30th October 83, the first record of this species for the
Isle of May. Unfortunately it had flown away before I returned
with my drawing equipment. Above is a detail drawing of a Grey
Phalarope which I drew on Fair Isle from a bird washed up
dead due to exhaustion.

This vicous bill, more reminicent of a falcon's, drew blood from my finger in a moment of carelessness.

Great Grey Shrike
Lanus excubitor

I caught it in the Top Trap on the 5th of May 83 and sketched it quickly in the Ringing Hut before releasing it. A quick sketch with colour notes was sufficient for me to work up later. It stayed till the 7th preying on the other migrants. I couldn't find it's larder - they collect prey impaling them on thorns etc, hence their nick-name - 'butcher-bird.' Other migrants present included Brambling, Goldfinch, Spotted Flycatcher, Lesser Whitethroat & Black Redstart.

Just a bit bigger than life size

upper tail pattern

genuine shrike dropping - luckily it missed the centre of the page!

diagnostic white patches
on tail (upper)

$\frac{1}{1}$

2nd Yr ♂ or ♀ Red-breasted Flycatcher
                    Ficedula parva
I caught this in the Low Trap today, the 31st May.
Oddly I caught another on the island 3 years
previously on the 31st of May 1980.

frequent tail flicking

Acrocephalus palustris $\frac{1}{1}$

olive green

light buff green   buff   legs:
                                    pale brown buff

Marsh Warbler, Low Trap, 21st June 83, they
are only really distinguishable from their close relative the
Reed Warbler by measuring wing formula. Another Marsh
Warbler had been caught and ringed on the 4th June.
This is a 'lifer', the first time I have ever seen this species!

♀

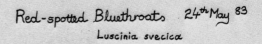

♂ with beautiful blue gorget.

I caught them both in a single shelf mist net in the ditch by the High Road

all ¹⁄₁

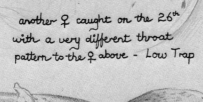

another ♀ caught on the 26th with a very different throat pattern to the ♀ above - Low Trap

28th May
¹⁄₁
♀ Pied Flycatcher, this one was in a bad way and died later. It was trying to feed amongst the kelp and rocks in Pilgrim's Haven but the weather was too bad. The last few days the island has been battered by very cold northerly force 7-8 winds with driving rain most of the time.

♂ Blackcap ¹⁄₁
caught in Top Trap   29th May 83

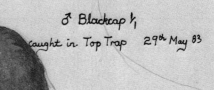

caught later on in the Low Trap

♀

1st year ♂ Redstart, life size detail
painted from a live bird held
briefly in the hand.

more buff on back

warm
brown & buff

1st May, 83

1/1

eye orange ring / dull red skin

grey mixed pinky brown.
along top crown

dull red along gape

dull grey (whitely bk)

grey around
cheek/white

warm
buff brown.

warm ochre/buff brown
patch on throat

pinky brown

Turtle Dove, I dazzled this one at night and
sketched it's head detail before ringing and releasing it
in the morning, 14th May.

lovely lemon
yellow underwing coverts

plumage detail of a
♀ Brambling found dead (emaciated)
on the 1st May, painted on 2nd

smaller than life size

¹⁄₁

(Gowk) Juvenile Cuckoo, 18ᵗʰ July 83, which
I caught in the Arnott Trap, sketched at the
Low Light.

bright red gape

Short yellowish pink legs

long beautifully marked tail

feathers fluffed up making them
hardly recognisable as birds, their
heads well tucked in.

Many roost in groups such as these birds
which I sketched by torchlight in a rock
crevice. Some singles were roosting on
top of dead nettle stems in ridiculously
exposed sites

Goldcrests , 30ᵗʰ September 83

♀ ½/₁

unfortunately many succumb
to exhaustion, the rain and cold wind

♂ ½/₁

prominant gold/brown
shoulder patch

hunched jizz due to the cold

♀ lacks the red centre
to the crown

tarsus warm brown
(yellowy) soles)

♂ Firecrest   Regulus ignicapillus ×⁵⁄₄,
plumage detail of a male present on the island
from the 1ˢᵗ to 7ᵗʰ April 1983 (caught in the Top Trap
on the 2ⁿᵈ). It arrived during a N·E force 2 to 4
wind, other migrants present included a woodcock
and the first two wheatears recorded this year.

not as bright an individual as
the ♂ I've seen previously

details from bird 'in the hand'
painted up later

fluffed up due to the cold weather

c ½/1

dullish crown stripe, basal ⅓
of lateral crown-stripe
darker than rest.

but dull &
orangy underside

slightly g edge to dull db/g.

diagnostic rump

Pallas's Warbler   Phylloscopus proregulus
14th November, 1983
A rare vagrant from Asia which was spotted by
Dave Pullan around midday feeding on the cliffs at
the South Ness. Later we caught it in a mist net
after a few attempts, it was so small that it
twice flew through the mesh of the net.

ring no. 9H5521   wg 50.5mm, weight 4.2 grammes

it was catching many flies
on the wing, the rump marking
very prominent in flight.

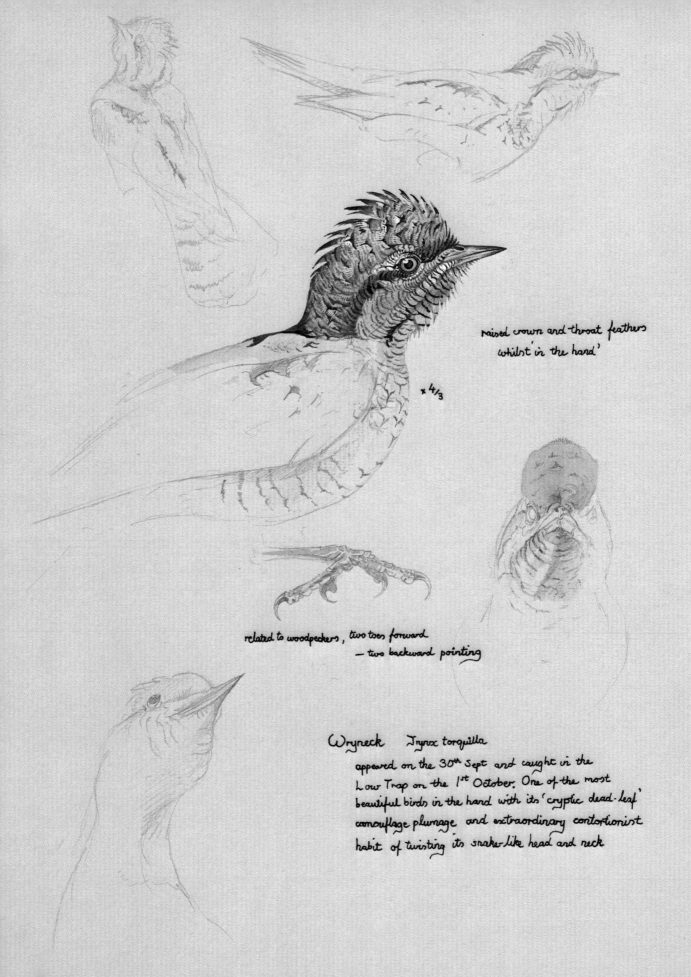

raised crown and throat feathers
whilst 'in the hand'

x 4/3

related to woodpeckers, two toes forward
— two backward pointing

Wryneck   Jynx torquilla
appeared on the 30th Sept and caught in the
Low Trap on the 1st October. One of the most
beautiful birds in the hand with its 'cryptic dead-leaf'
camouflage plumage and extraordinary contortionist
habit of twisting its snake-like head and neck

A contrast in plumages and lifestyle, the garish
Rose-coloured Starling (Sturnus roseus) a rare vagrant
from the grassy steppes of Asia. An adult ♂ was present
on the island on the 2nd June 83, its plumage matching
the pink clumps of thrift through which it was foraging.
Compare the camouflaged plumage of a Wryneck, a
reticent woodland bird. Both painted life-size from the
skin collection of the Royal Scottish Museum, Edinburgh.

lemon y

dull brown/green

gray white

blueish bill

white

Siskin ♀ 1/1

rufous

Ochre eyestripe
& white

blueish (yellow gape)

warm brown

Reed Bunting 1/1

ochre/b. blue black

black

deep
yellow

orange
whiter under bill

Brambling 3♂

Ring Ouzel, this bird I caught
in 'Elsie', the outside toilet building! I
was walking by and heard it flapping
about inside so I caught it by hand.

Some drawings of migrants from a big fall of
Scandinavian migrants which arrived today, 29th September.

93

lemon yellow orbital ring
set close

brown

black/brown

brown shading

Fieldfare, 1st year ♀ , ½
from a bird in the hand.

Redwing adult ½

21st October 83, some migrant thrushes are passing
through today but most are flying straight onto the
mainland despite the westerly winds. Blackbirds and
Song Thrush in smaller numbers, the lack of easterly
winds has curtailed the arrival of huge thrush flocks
usually associated with the latter ½ of October.

Long-eared Owl details   13th November 1983

There are at least 6 of these owls on the island
and we caught 3 roosting in the traps

Long-eared Owl, 15ᵗʰ Nov, roosting in a crack
in the cliff above Mill Door

eye to eye! a Short-eared Owl hidden in
the long grass above Horse Hole, 22nd Sept

Kestrel 24th Sept, sketched from
the Low Light window as it hovered
against a strong easterly wind. I later
watched it catch a mouse.

owl pellet containing
mostly mouse remains

25th Sept

Short-eared Owl sitting on the cliffs at Tarbet Hole getting
some respite from the party of birdwatchers combing the plateaux
in search of an elusive Quail. The other owl was flapping
about over the sea ½ mile off the island mobbed by lots
of gulls.

tail feather

fawn brown

♀ or imm Kestrel
sketched from the window of the Low Light
as it hovered against a force 6 northerly wind
10th September

juvenile Wheatear ½ 10th Sept
plumage details from a
bird in the hand.

22.9

broad buff tips.

When the Kestrel appeared this was the only piece of paper
to hand as I had just finished drawing the Wheatear. So the
Kestrel sketches grew round the page as I didn't expect to
get so much done, normally they don't remain in the one area
for long.

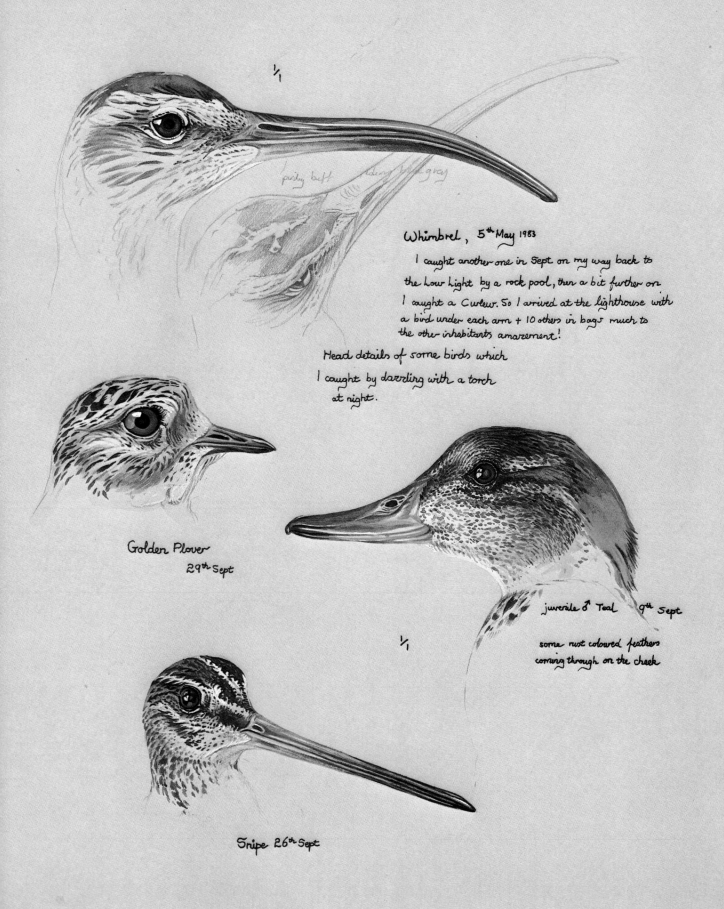

1/1

puffy buff   dusny pink gray

Whimbrel, 5th May 1983

I caught another one in Sept on my way back to
the Low Light by a rock pool, then a bit further on
I caught a Curlew. So I arrived at the lighthouse with
a bird under each arm + 10 others in bags much to
the other inhabitants amazement!

Head details of some birds which
I caught by dazzling with a torch
at night.

Golden Plover
29th Sept

juvenile ♂ Teal   9th Sept

some rust coloured feathers
coming through on the cheek

1/1

Snipe 26th Sept

99

upperwing & tail

@ ⅔ life size

Juvenile ♀ Lapwing , 6-7ᵗʰ September 83
Found by Rab Morton near the North Horn, I presume it was
suffering from botulism as it couldn't move it's wings or legs
at all — it was too far gone to survive. Gorgeous iridescent
plumage in strong light

Purple Sandpipers

♂

♂   ♀

juvenile with buff tips to
the wing coverts

Turnstones, some still in full summer plumage
others moulting into their drabber winter
plumage

♀

Purps and Turnstones at roost on Foreigners' Point   7th Sept 83.

101

Birds moult to renew their feathers each year as they become bleached and abraded. In the case of waders such as the Turnstone below they moult their primaries (main flight feathers on wing) upon arrival from their breeding grounds in Canada, Greenland or Scandinavia. They moult them in sequence from the inside outwards a few at a time so as not to impair their flying capabilities

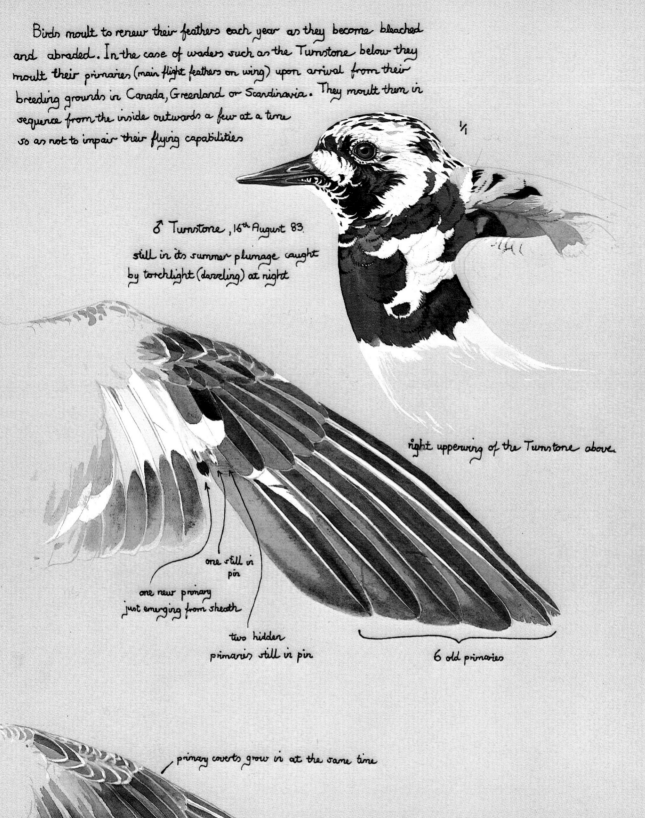

1/1

♂ Turnstone, 16th August 83.
still in its summer plumage caught
by torchlight (dazzling) at night

right upperwing of the Turnstone above.

one still in
pin

one new primary
just emerging from sheath

two hidden
primaries still in pin

6 old primaries

primary coverts grow in at the same time

Purple Sandpiper wing (moulting) 1/1, this individual had only one
tail feather left in !

2 faded and worn primaries soon to
be moulted and replaced

most of secondaries
missing

6 new primaries

one 3/4 grown + one still in its sheath not yet visible

resting amongst
bladder wrack

sleek diving jizz

surfacing amongst seaweed

½ asleep

with yellow gape

fine light buff flecks + dark not f

off W rump

dark brown top & tail

Little Grebe 27ᵗʰ September 83
feeding and resting amongst the seaweed
in the channel at East Tarbet.

peering under a stone

rubbing oil onto its back
from the preen gland

up-ending

looking right
under for food

Greenshank, this juvenile was present feeding
along the shores of the loch from the
1st – 7th Sept 83

very difficult to sketch as it bobbed up and down in
the waves and between dives.

a ♂ Long-tailed Duck in transitional plumage in relatively
sheltered water off the South Ness (wind = SW force 6 to 7)
29th October 83

Curlew and Redshank at roost
Foreigners' Point

30th September, 83

by the 20th of October the bill has
darkened, the white cheek patch is
beginning to appear and the flanks
are getting whiter

20.10

occasionally it gave
a throw-back head display

20.10½ asleep

Immature (1st Year) ♂ Goldeneye, almost
resident on the Loch

wing stretching

106

¹⁄₁

from above

Storm Petrel (life size) caught in a net by the Low Light on
the night of the 5ᵗʰ August by Bernie Zonfrillo. It was attracted
to the net by tape recordings of their 'purring' song calls. One
of the 4 he caught had been ringed at Noss Head, Wick,
Caithness 9 days earlier (255 kms in a straight line).

bluey tinge to
back F.

tail from side

@ 2/3 life size

Juvenile Gannet   19th September 1983
found very weak and with a damaged wing on the landing
at Altarstanes, unfortunately it did not survive. It probably
came from the nearby Bass Rock (10 miles away), a victim of
the SW gales of the past few days.

underside

juvenile Gannet details   Sula bassana (named after the nearby)
     21st September 83   Bass Rock

wing length = 483 mm

@ ½ life size

posture when entering
water during plunge dive
for fish

Adult Gannet, 5 years +
washed up freshly dead on Silver Sands
today, the 27th September 83. This one
had lemon yellow coloured eyes.
wg = 487 mm

@ ¼ life size   pale 'winter' head.

tail in moult, new feather
+ one coming in, rest very abraded.

lovely foot markings
@ ⅓ life size

new

new

P5 in pin

P9 at
stage 4.

110

getting some respite from the mobbing gulls on
the rocks at low tide on the North Ness, kelp in
the background

Bonxie, 17th June 83
This Great Skua arrived on the 12th and has
been causing havoc amongst the gulls whilst making
sorties to steal gull chicks to eat.

a Herring Gull mobbing the skua, usually the
tables are turned with the skua harrassing the
gulls making them regurgitate food to lighten
themselves for an escape

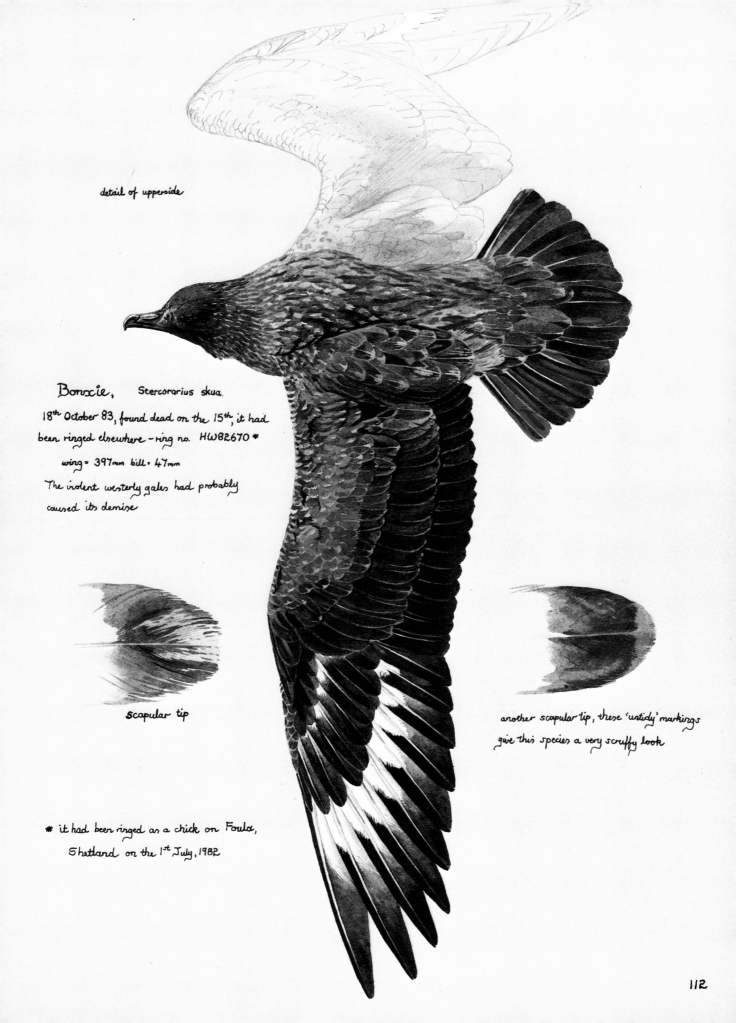

detail of upperside

Bonxie, Stercorarius skua.
18ᵗʰ October 83, found dead on the 15ᵗʰ, it had
been ringed elsewhere – ring no. HW82670 *

wing = 397mm bill = 47mm
The violent westerly gales had probably
caused its demise

scapular tip

another scapular tip, these 'untidy' markings
give this species a very scruffy look

* it had been ringed as a chick on Foula,
Shetland on the 1ˢᵗ July, 1982

an adult and a first year
Cormorant drying their wings
on the North Ness, 24ᵗʰ February,
a very foggy day.

alarmed by the blasts of depth
charges dropped by a distant
naval vessel.

these breed on an island c 18kms
away – the Lamb.

blue/black

tail in moult

Adult Cormorants in
breeding plumage at roost on
the cliff near Greengates, 1ˢᵗ March 83

113

adult Puffin in winter plumage

adult, winter plumage.

Razorbills

1st year

'Little Grebe' like jizz

Little Auks

5th February 83, sketches of some auks feeding off the South Ness sheltered from the NNW force 7~8 wind. Luckily up to 7 Little Auks were just offshore affording excellent views of this species which normally stays well out to sea. A Puffin, normally well out to sea during the winter, was diving close inshore. All the auks displayed a very hunched jizz between dives with wings drooping. One Little Auk was diving only 2 metres from me at times, when relaxed these tiny auks had the jizz of a Little Grebe.

6th February, 76 Little Auks seen in a 2 hour sea-watch off the South Ness. Also 2 Iceland Gulls (adult & 3rd Yr) and a 1st year Glaucous Gull amongst the roosting Herring and Greater Black-backed Gulls off the Maidens.

4
GREY SEALS

The grey or Atlantic seal (*Halichoerus grypus*) is a traditional breeder on the Isle of May. Records as far back as 1508 show that seals were killed here, and one account relates that in March 1508 the sum of thirteen shillings was paid 'to the heremyt of Maii that brocht ane selch to the King'. ('Selch' is an old fisherman's name for the seal and 'heremyt' refers to the hermit staying on the island at that time.) The May is one of only three breeding rookeries of grey seals on the east coast of Britain, the others being the Farne Islands, Northumberland, and Scroby Sands, Norfolk. Common seals (*Phoca vitulina*) occasionally visit the island but they generally prefer sandy estuaries and tidal sandbanks such as the Tay estuary: the only ones recorded on the May were two which I saw hauled out on Silver Sands on 2 June 1983. Common seals are sometimes confused with yearling grey seals but in fact their snub noses are quite different from the grey's 'roman' nose profile. There were no recent grey seal breeding records until 1956 when at least one pup was born. In the last few years, however, there has been a dramatic increase in breeding and 615 pups were born in 1982 and 350 pups in 1983. This increase coincided with the disturbance of seals on the Farne Islands from 1977 onwards. There used to be over 2000 breeding cows on the Farnes but they were causing severe erosion to the puffin breeding areas on some of the islands in the group and so bird scarers (and humans) were used to keep the cows off the islands at the critical pupping time. As a result some of the cows moved north to the May.

Tagging and other methods of marking such as radio telemetry are being carried out on the island by the Sea Mammals Research Unit from Cambridge University which studies and monitors the number of seals around the coast of Britain. Tagging has shown that the yearlings can travel great distances in a short time. One male pup, two to three days old, which was tagged on 23 November 1959 on Brownsman, one of the Farne Islands group, was then caught and released on the May on 21 December when only some thirty days old. Nine days later a twelve-year-old boy found this same pup freshly dead in his fishing nets at Kvalavag, Karmy, in Norway. It was a remarkable swim for such a youngster. Most of the pups are in fact recovered on the east coast of Britain.

The plastic rototag (actual size) is fitted to the webbing of one of the
rear flippers. A reward is given upon sending relevant details to
the address on the underside of the tag i.e. 'inform London Zoo'

During the spring and summer the number of seals rarely exceeds fifty
to a hundred, but it starts increasing from early October. By the end of
October the first pups are being born, although the peak pupping time
occurs in the first three weeks of November. I have only seen three births
actually taking place; they are usually so rapid, taking only a few seconds,
that they are easily missed. Many seem to occur during the hours of darkness. The pup at birth is about 36 inches long, weighs just over 30 lbs, and
retains part of its umbilical cord for a few days. At this new-born stage the
skin is wrinkled loosely round the body but this quickly fills out as the
pup is fattened on the mother's rich milk – which is some fourteen times
richer than cows' milk. The cow suckles the pup usually three to four times
a day, for two to three weeks, before deserting it. Some cows lie up with
their pups during the day but the majority spend most of this time in the
sea, only coming ashore to suckle. Apparently the cows do not eat during
this three-week period. This is borne out by their lean shape, the pelvic
girdle clearly visible beneath the skin, by the time they leave their pups.
After three weeks the pup has trebled in weight and has started to moult
out of its puppy coat. This usually takes four to five days with the head and
flippers moulting first. After this the pups start to explore the seashore and
rock pools appearing to revel in their new environment. Some of them
however, don't make it to this stage: storms take their toll, some pups are
stillborn, some get accidentally crushed by adults, and malnutrition is a
further hazard.

With their large saucer-shaped eyes and surrounding wet patch caused by 'tears' the pups are very appealing. They spend most of their time sleeping, rolling about on their backs, stretching and scratching their flippers. Many appear to be contentedly dreaming. They can, however, be very vocal, making soft bawling and groaning noises. When hungry and looking for its mother a pup's bawling becomes even louder and more incessant. If disturbed, seals make menacing, snarling noises and usually move forward to try and bite one. They have teeth from birth and, given the chance, could inflict a nasty wound. Some of the cows are very protective to their offspring and fearlessly charge at any human intruder. On land they can have quite a turn of speed for an apparently cumbersome creature.

A totally blind cow pupped on the rocks less than thirty feet behind my hide on the other side of the bridge path. One could approach her quietly downwind with ease, her glassy, opaque eyes quite unseeing. Despite this handicap she managed to raise a healthy pup. Most of the time she lay in a pool just below her pup; sometimes she was hardly visible as she slept underneath the floating seaweed and driftwood. Yet she managed to find her pup by scent without apparent difficulty. I presume she was able to locate food with the aid of her sensitive whiskers. She kept away from all the other cows in the colony as she probably couldn't cope with aggression from other females. Still she did have a bull all to herself! Whilst out 'dazzling' birds at night I have come across a few other blind seals, one or both eyes affected by glaucoma.

The mature bulls are larger than the cows, their head-profile is uglier and more convex and their necks are thicker and more wrinkled. In the rookery they are outnumbered some five to ten times by the cows. They establish territories within the colony which they vigorously defend against any intruding bulls. Real fights between bulls are few and far between, most are show-downs in which the weaker of the participants backs off after making threatening growls. Some bulls may pursue the intruder for a short distance, but staying within their territory. Many of the bulls are scarred especially around the neck area, some the result of wounds inflicted by cows defending their pups from the lumbering males. Usually a cow 'flippers' the bull to keep him at a distance, in other words she hits the bull with her fore-flipper in a rapid scratching action. If this fails to deter the bull she lashes out and bites him and he invariably backs away. The cows usually become receptive to the bulls when their pups are over two weeks old. Mating takes place in the water or on land and can sometimes seem very brutal as the bull grips the cow's neck in his teeth. After a few weeks of territorial aggression and copulation, without eating, the bull is literally a shadow of his former self.

The seals are creatures of contrast. In the sea they are fast and agile swimmers with a sleek grey coat lighter on the underside with darker blotches. Individuals can be recognised by the pattern of the blotches, some being much more heavily marked than others. They sleep underwater and can reputedly stay under for twenty minutes at a time. Normally, however, one sees them 'bottling', that is sleeping vertically in the sea with just their nose or head above the surface. On land they are slow, cumbersome movers, manoeuvring themselves along with the aid of their fore-flippers. Before the seals dry out there is a marked border between the wet and dry fur. When their fur does dry out they are mainly brown in colouring with the blotches not so prominent.

Two immature seals on the island this year had pieces of a nylon fishing net wrapped tightly round their necks that were biting deep into the flesh and causing suppuration. Obviously the seals had become ensnared in nets which they were raiding and had chewed themselves free after much twisting. Probably the seals would eventually die but there was no way to catch them to remove the netting.

In order to draw the seals in as relaxed a manner as possible (for both myself and the seals!) I decided to build a temporary hide. The most suitable place in many respects was by the Iron Bridge. I built the hide out of old driftwood up against the wall of the bridge in September, well before the first seals came ashore. This site afforded some protection from the predominantly westerly winds and salt spray; moreover it was not silhouetted against the skyline. The hide overlooked a gully which was flooded at high tide and at low tide a large pool was left amongst the rocks and seaweed, and I was able to sketch seals swimming and sleeping in the water. The surrounding area of rock and vegetation had, at any one time, between six and twelve cows with their pups. The northerly aspect meant that I had good light during the whole of the short winter days. More important, though, I could get into the hide from the path without causing undue disturbance to any other seals. By using a relatively small group of seals I had the advantage of being able to get on more intimate terms with them. For example, in order to get details and different angles of the bull to finish previous sketches, I had to observe the same animal over a period of time.

young pups

day old pup with part of umbilical cord
still present, also wrinkled body

Nov 1982

115

very young seal pups, Rona Nov 82

dilated nostrils on inhalation
whilst sleeping

Pilgrims 24.10

117

asleep on the boulder beach

pup 4~6 days old   Pilgrims Haven
24th October 1983

118

pup ½ way through moult

fat 3 week old pup just
starting moult

Rona   23rd Nov 83

--- in moult

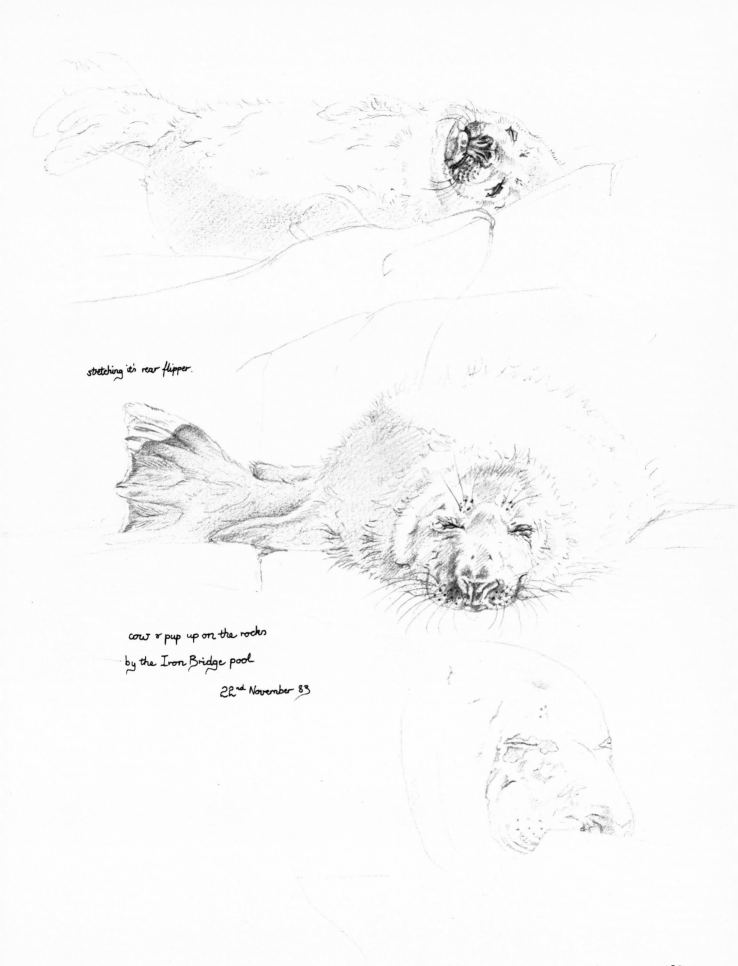

stretching it's rear flipper.

cow & pup up on the rocks

by the Iron Bridge pool

22nd November 83

young pup, Pilgrims Haven
24th October '83

121

head detail of a young pup, Rona, Nov 82

Rona 28.12.82
on algae covered rock

sleeping grey seal pups (fully moulted)

amongst rocks and driftwood
Pilgrims Haven, 29.12.82

tufts of natal coat

remainder of the body which
I didn't have room for up top

a fine fat pup, just about fully moulted out with
only a few tufts of its' natal coat left on its back,
most of its time has been spent exploring the pool and
rolling amongst the kelp — 20th November 1983.

124

Grey Seal pup @ 3 weeks old, very fat, moulting out
it's white 'puppy fur' for a sleek new pelage

Silver Sands, 24th December 1982

125

North Ness   24.11.83
Moulted pup chasing a young eider duck round and round
a stagnant pool, the eider eventually got bored and waddled
back to the sea. This pup was having great fun exploring
the pool and playing with a tyre and a small oil drum
floating in the murky green water.

cow lying up with her pup, she'll soon
leave the pup as it is fully weaned and she
has lost a lot of weight

½ asleep in the water, part
of her back dried out

from the Iron Bridge hide, 17th November 83

Seal cow with her pup by the Iron Bridge pool, 15th Nov 83, she
has lost the outermost claw on her left fore-flipper.

♀ in Pilgrims' Haven
7th Nov

Moulting pup, Rona   8th Nov 82

129

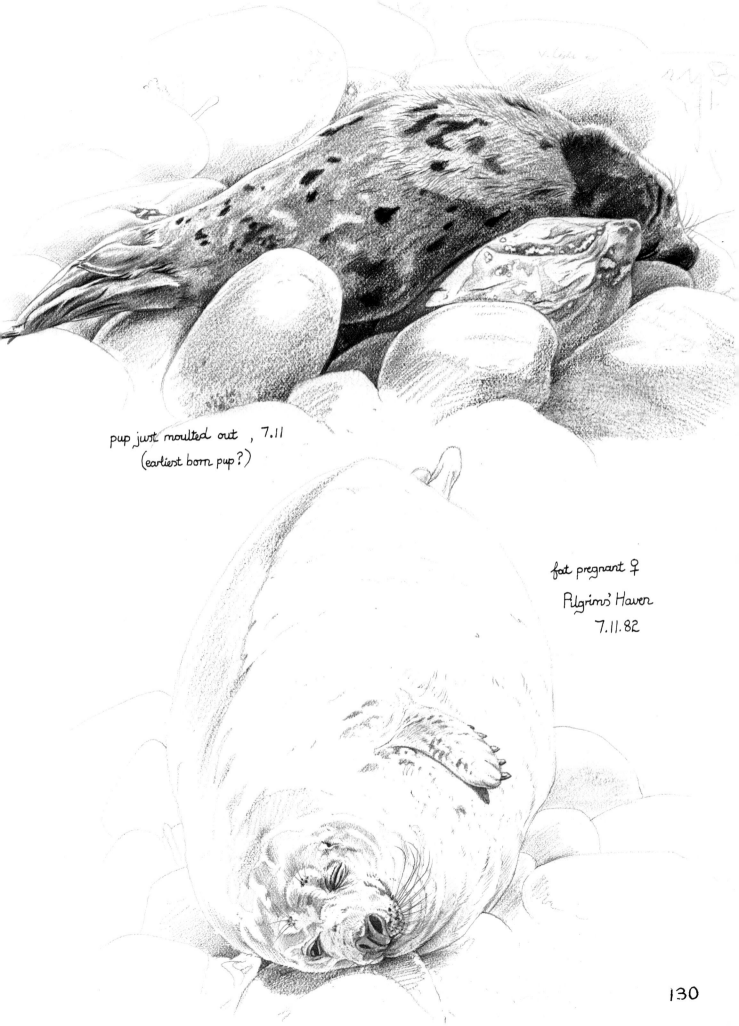

pup just moulted out , 7.11
(earliest born pup?)

fat pregnant ♀
Pilgrims' Haven
7.11.82

130

A cow with her pup suckling, Iron Bridge Pool 23rd November, 1983.
I sketched this from a few feet away while the pup fed, she kept
a watchful eye on me. She was an agressive cow and not averse to
charging at any intruder, on her rear flipper was a red 'streamer' with
a tag no. 24944 (marked on the island in 1982)

this cow was totally blind but managed to bring
up her pup successfully, she was under the Iron Bridge
away from all the other cows.

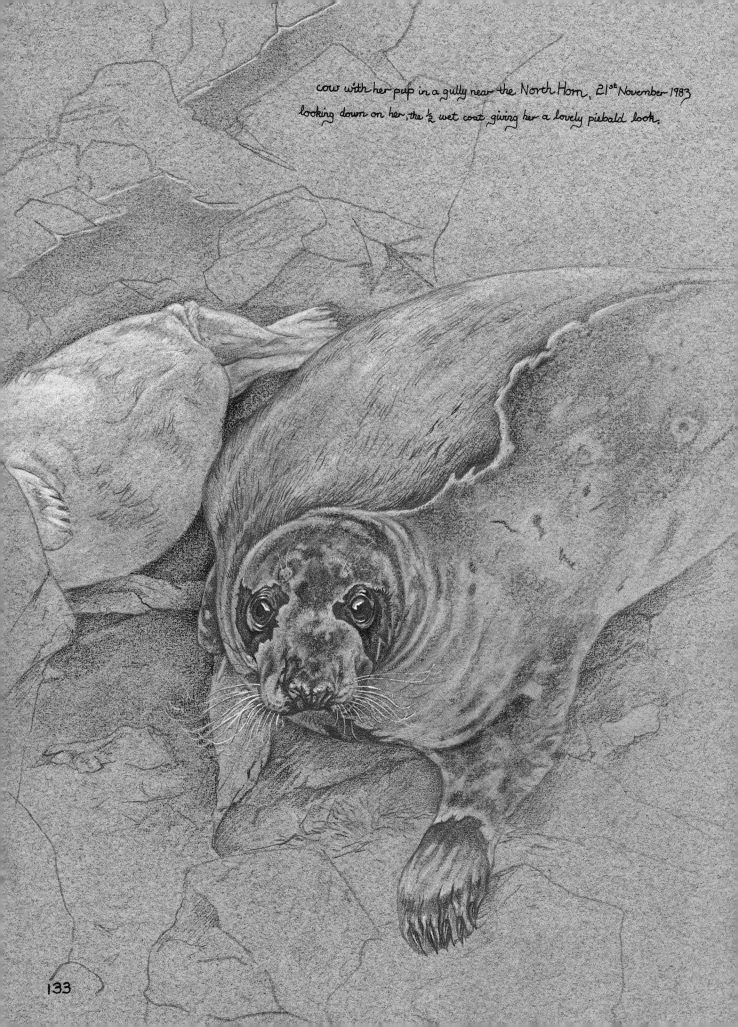

cow with her pup in a gully near the North Horn, 21st November 1983

looking down on her, the ½ wet coat giving her a lovely piebald look.

133

Yearling Grey Seal, North Ness 21ˢᵗ October 83
hauled up on the rocks, very weak with bare patches of
skin round the head (mange?). I sketched it from 2 feet
away as it snarled at me.

134

Kirkhaven 28th Oct 83

28.10

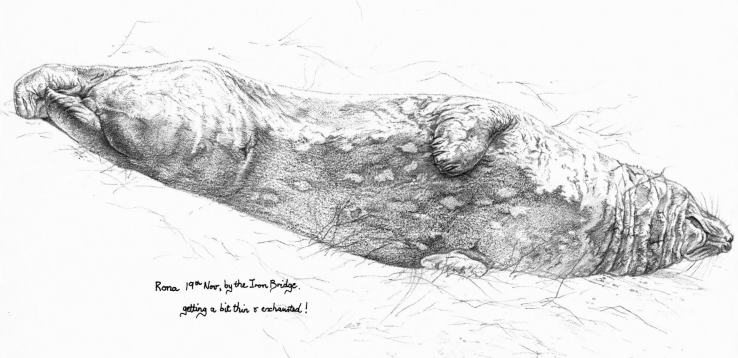

Rona 19th Nov, by the Iron Bridge.

getting a bit thin & exhausted !

Grey Seal Bulls at rest

135

A bull threatening another which is
venturing too close to this ones territory. Prolonged
fights seldom occur between bulls, most of the
damage is done to them by belligerent cows.

Iron Bridge pool, 22ⁿᵈ Nov 83

bull and cow asleep in a pool,

Iron Bridge 18th November 1983

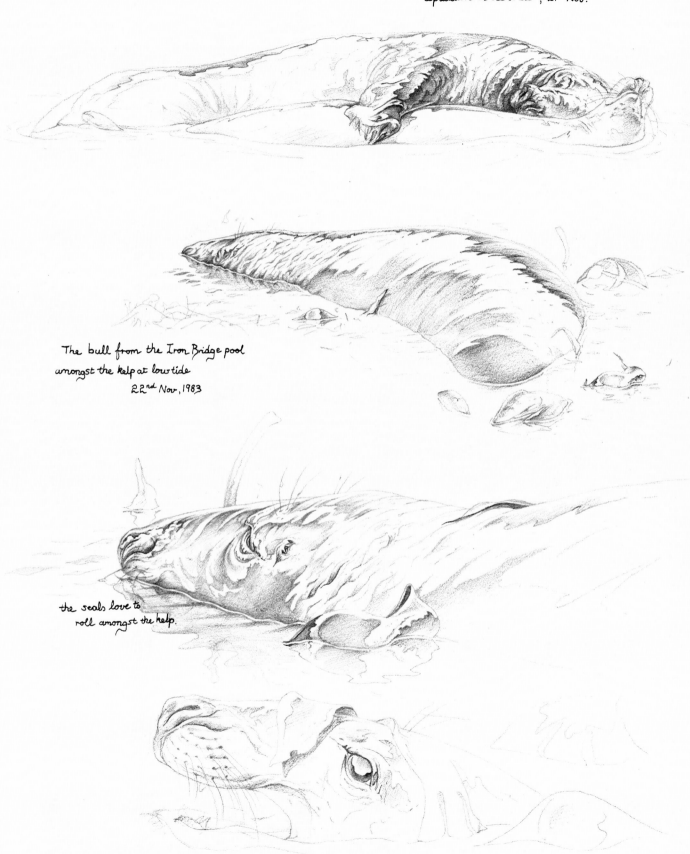

copulation in the water, 21st Nov.

The bull from the Iron Bridge pool
amongst the kelp at low tide
22nd Nov, 1983

the seals love to
roll amongst the kelp.

Bull seal in the pool by the Iron Bridge
20-23rd November 83

bottling

The bulls are easily distinguished from the cows by their larger size, more convex profile, wrinkled neck and more ugly face.

surfacing for air

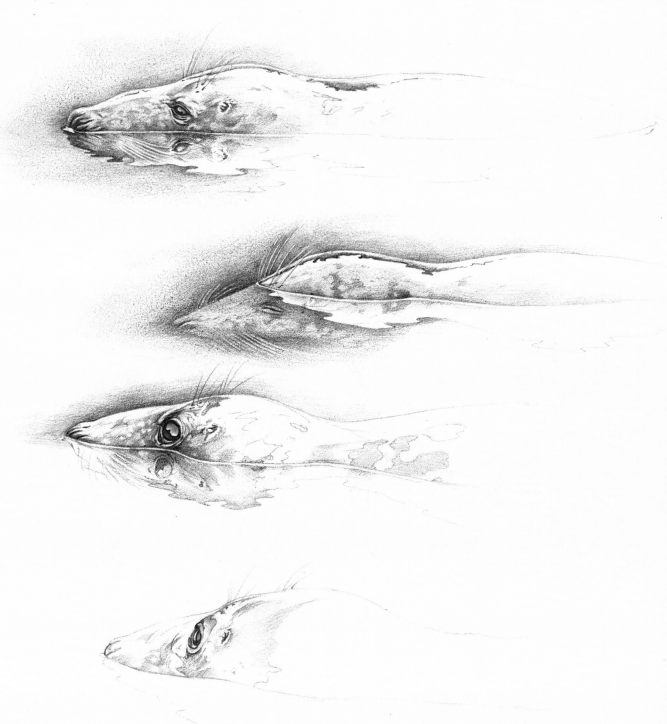

grey seal cows floating 'crocodile-like' in a pool by the Iron Bridge, 18th November 83

♀♀ sleeping in the Iron Bridge Pool with reflections of the ripples
running across their bodies , 15th November 1983

♀ in IronBridge pool, Nov 83

142

pool in shadow, dead dock stems in the
water (deep brown coloured)

Rona 16th November 83

Cow wallowing in a stagnant pool a few feet away
from me, it's pup lying asleep beside me. She had a
dry patch of fur contrasting with the sleek wet coat.

143

Grey seals asleep in the sea with heads
above the surface, called "bottling." Rona 27.10.83.
Some sleep below the surface and can stay submerged for
up to 20 minutes!

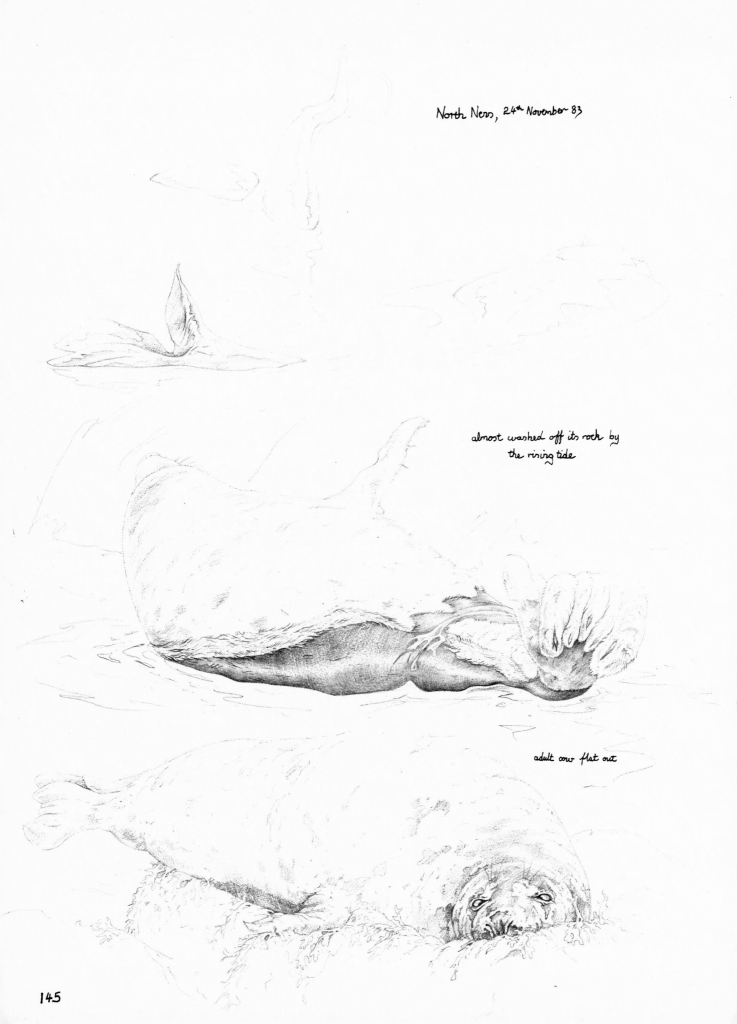

North Ness, 24th November 83

almost washed off its rock by
the rising tide

adult cow flat out

'loafing' cows up on rocks on the North Ness
at low tide   24th November 83
(whilst sketching the seals a Black Redstart landed
one foot away from me till it realised I wasn't a
rock although I felt like one in the freezing wind!)

resting cows on Rona, 16th November 1983

♀♀

♂

Sleeping seals, North Ness 21ˢᵗ Oct 83

148

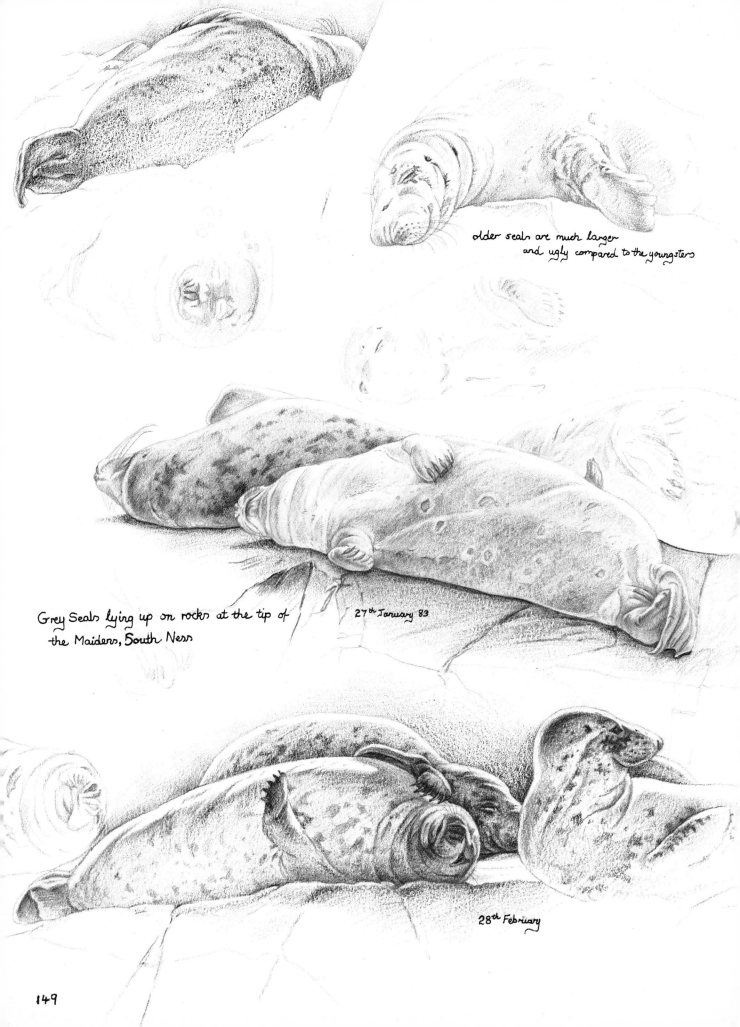

older seals are much larger
and ugly compared to the youngsters

Grey Seals lying up on rocks at the tip of
the Maidens, South Ness

27th January 83

28th February

149

Rear view of a cow just hauled out of the water,
pool by the Iron Bridge, 21st November 1983.